Mapping and Topographic Drafting

John D. Bies
Human Resources Development Consultants
Memphis, Tennessee

Robert A. Long
Drafting and Design Technology
Mississippi State University
Mississippi State, Mississippi

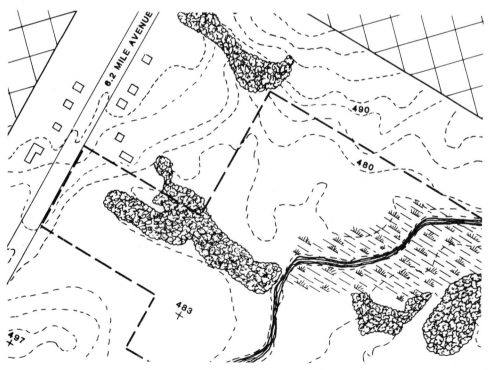

Published by

IE33   **SOUTH-WESTERN PUBLISHING CO.**

CINCINNATI   WEST CHICAGO, ILL.   DALLAS   PELHAM MANOR, N.Y.   PALO ALTO, CALIF.

ISBN: 0-538-33330-8

Library of Congress
Catalog Card Number: 82-62044

2 3 4 5 6 7 D 9 8 7 6 5 4
Printed in the United States of America

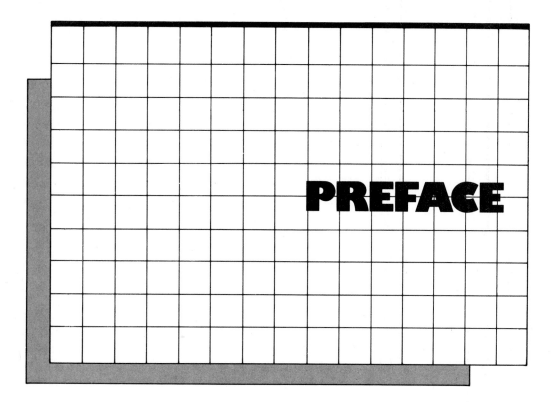

PREFACE

In MAPPING AND TOPOGRAPHIC DRAFTING, the authors have developed a professional level text designed to meet the needs of students who require training in this area as part of course sequences in cartography, geography, civil technology, civil engineering, surveying or allied areas. Throughout the text, the authors have maintained consistence with all relevant professional standards.

A comprehensive approach, this book includes chapter-length material unavailable in any other text presentation on surveying, celestials, career opportunities in the field, and applications of Computer Aided Drafting.

Other topics include Mapping and Topographic Instruments and Materials, Topographic Drawings, Conventions and Practices, Contours, Traverses, Profiles, Photogrammetry, and Reproduction Techniques.

All areas of basic theory and field application are reinforced with a thorough illustration program including professionally drawn topographic maps and detailed sections. Each chapter includes a summary, a listing of key terms, and appropriate review questions and activities.

Requisite mathematics does not exceed right angle trigonometry and is included only where necessary for the development of the subject. Problems are carefully worked out in a step-by-step fashion including relevant illustrations to back up the mathematical procedures.

Special effort has been made to publish MAPPING AND TOPOGRAPHIC DRAFTING to the latest professional standards in the field including:

Photogrammetry: Treatment of specialties within the field such as Radargrammetry, Hologrammetry and Space or Satellite Photogrammetry and accompanying photographic material emphasize the currentness of the treatment.

Topographic Drawings: The heart of the book, this material deals with use of hand drawn maps and available published data for use in the field by geographers and engineers.

Computerized Graphics: Featuring the applications of modern digital equipment to a field situation, the systems are thoroughly described and system interaction brought to "state of the art."

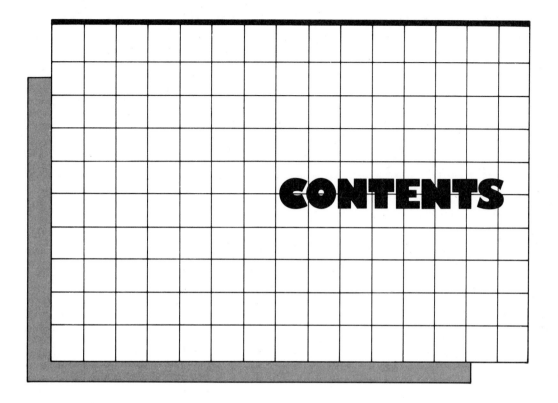

CONTENTS

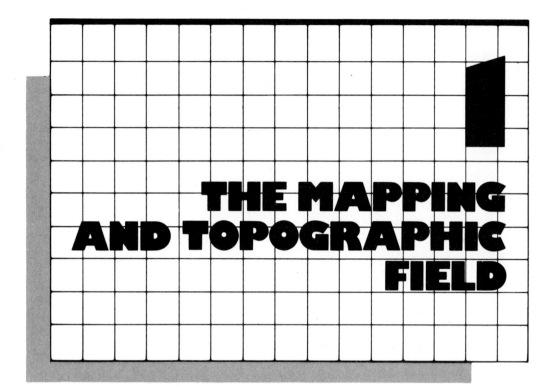

THE MAPPING AND TOPOGRAPHIC FIELD

Topographic mapping is the process of recording the surface features found in a given area or region of the earth. The topography of the earth's surface can be drawn on either maps or charts. The term *chart* is reserved for maps upon which plottings will be made, such as aeronautical and nautical charts. Topographic drawings are often referred to as *topos* in the engineering and planning fields.

Topographic maps are drawn by drafters and cartographers who use a combination of instrument drafting and scribing techniques along with freehand drawing representations. Increasingly, topographic information is being fed into computers for processing and production of map plottings.

Topographic drawing is an interesting segment of the drafting and drawing field. Because of the techniques used, topographic drawing is considered to be neither an exact science nor a pure art form, but a combination of both. It is part science in that the topo drawer must rely on and report detailed and accurate information pertaining to surface bounds and characteristics. Topo drawing is a form of art in that it requires the design and reproduction of drawings that must present a large amount of information efficiently and effectively.

Small-scale mapping is performed mainly by governmental agencies. Large-scale mapping, on the other hand, is performed primarily in the private sector, by engineering and related firms. Private-sector maps are not published; they are reproduced in limited quantities.

In this book, emphasis is placed on private-sector drafting practices and techniques.

1.1 USES OF TOPOGRAPHIC DRAWINGS AND MAPS

Topographic drawings are used in many technical fields and professions. No single profession has an exclusive claim on their preparation and application. This feature of topos is a contrast to the "exclusivity" of architectural drawings, for example, and electrical plans. Every field or profession requiring the specification, planning, and positioning of tracts of land, natural or man-made objects, coastlines, or land areas, has a need for topo drawings and maps. Some of the common applications of topos and fields which use them are discussed in this section.

Surveying

Surveying is the process of determining the location, form, and/or boundaries of a tract of land by measuring its bounds (perimeter) and features. The data collected in a survey provide the essential source or base data from which topos are prepared. Without accurate surveys, many construction projects would be practically and physically impossible to complete, or would result in problems such as surface drainage or inadequate road design.

Surveyors or surveying firms are frequently contracted to survey tracts of land that are already recorded on existing to-pos. In such cases, additional surveys are conducted to compare present conditions to those recorded, or to subdivide a given tract of land. Another reason for surveying a second time might be to determine, on a finer grid or on a more detailed level, the characteristics of the land or positioning of stationary landmarks. Figure 1-1 is an example of two surveys that were conducted on the same piece of property in order to record two types of information.

Engineering

The engineering field (particularly civil engineering) relies heavily on the information provided in topo drawings. Surface topography is important in engineering projects such as those involved in construction of buildings, bridges, roads, harbors, airfields, dams, tunnels, and water supply and sewage systems. In some cases, the basic data concerning the physical characteristics of a project site can be obtained from local planning and assessor's offices in the form of topographic, soil, and property maps. However, if the information provided by these offices is not up to date or is not sufficiently detailed, the engineering firm will conduct a separate survey to provide a more accurate topo of the site. Sometimes the engineering firm automatically conducts its own survey, regardless of the scale and date of existing topos, as a precaution against possible error or miscalculation.

A complete topo of the site is part of any engineering or architectural project plan. As shown in Figure 1-2, the surface characteristics of the project site are indicated as well as the surface characteristics of land contiguous (next to) to the project site. Depending upon the project, the topo, along with the complete set of engineering drawings and specifications, must be reviewed

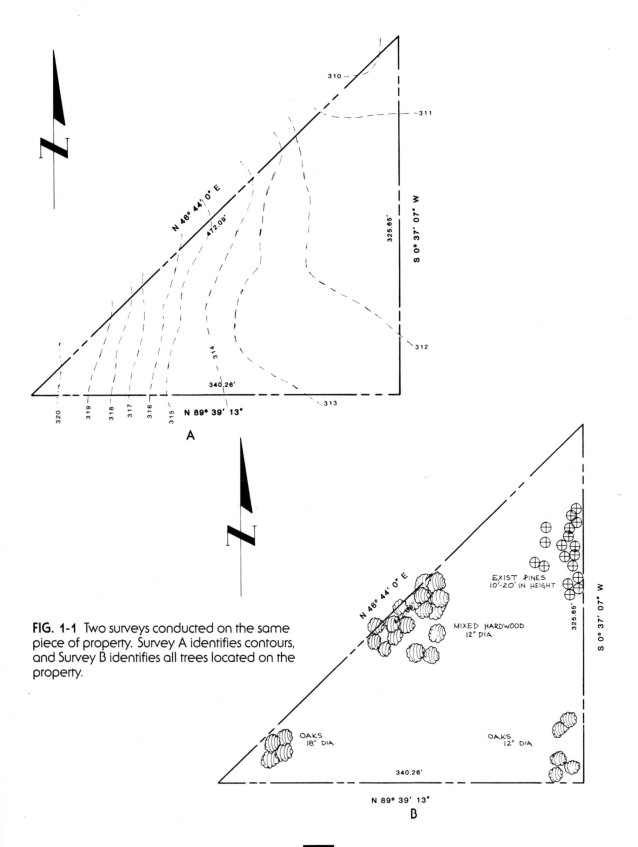

FIG. 1-1 Two surveys conducted on the same piece of property. Survey A identifies contours, and Survey B identifies all trees located on the property.

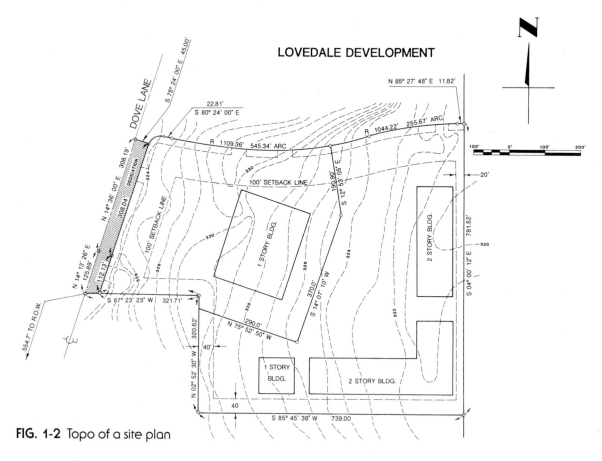

FIG. 1-2 Topo of a site plan

and approved by appropriate governmental agencies before construction can begin. Examples of these agencies are the engineering departments of the city, state, or federal government.

Hydrography

Hydrographic charts supply information pertaining to harbors, rivers, and other bodies of water. Information charts on hydrographic features such as navigational aids, underwater hazards, lights, and harbor facilities can be obtained from the National Ocean Survey and the Lake Survey Center. Publications of these organizations include small-craft charts, harbor charts, coast charts, general charts, and sailing charts. Figure 1-3 shows a hydrographic chart of a waterway.

Geodesy

Geodesy is a branch of science and mathematics that determines the exact position of figures, points, and areas along with the curvature of the earth's surface. An example of a topographic map based upon geodesic information is the geological map, Figure 1-4. Geological maps identify and locate geological features on topographic maps.

Geologic mapping is frequently accomplished for large surface areas (small-scale maps). Geodesic data are used to locate the exact locations of geological features such as general rock types, waterways, and rock structures. General geological maps can be obtained from the United States Geological Survey or from state geological

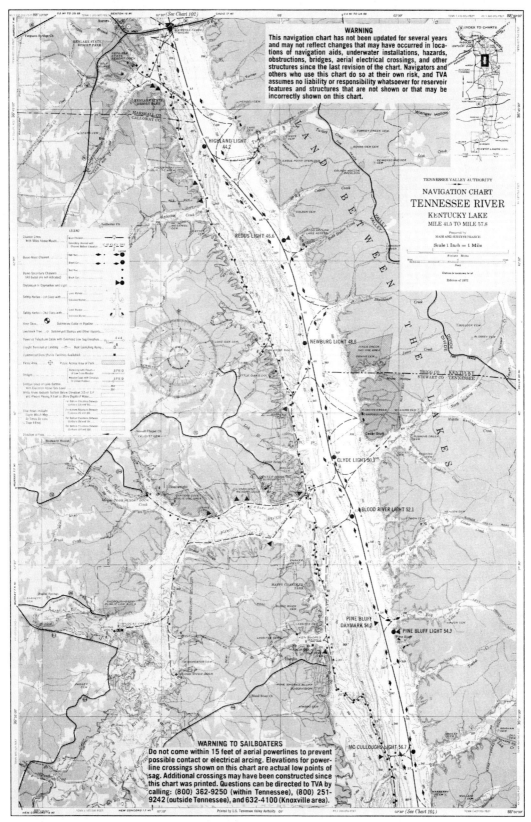

FIG. 1-3 Hydrographic chart of a waterway *(Courtesy of Tennessee Valley Authority)*

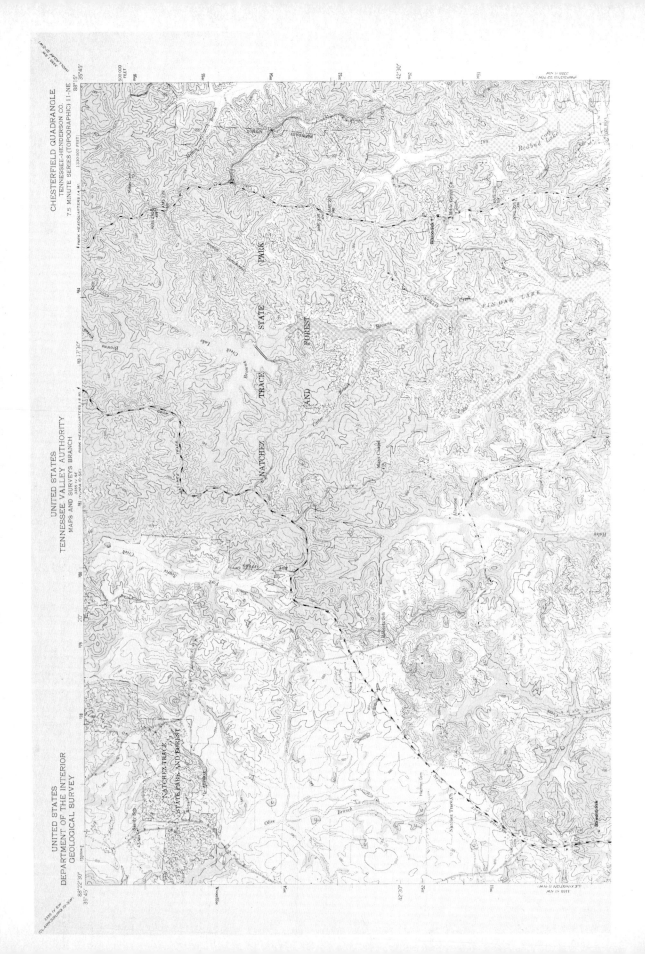

CHESTERFIELD QUADRANGLE
TENNESSEE-HENDERSON CO.
7.5 MINUTE SERIES (TOPOGRAPHIC) 11-NE

UNITED STATES
TENNESSEE VALLEY AUTHORITY
MAPS AND SURVEYS BRANCH

UNITED STATES
DEPARTMENT OF THE INTERIOR
GEOLOGICAL SURVEY

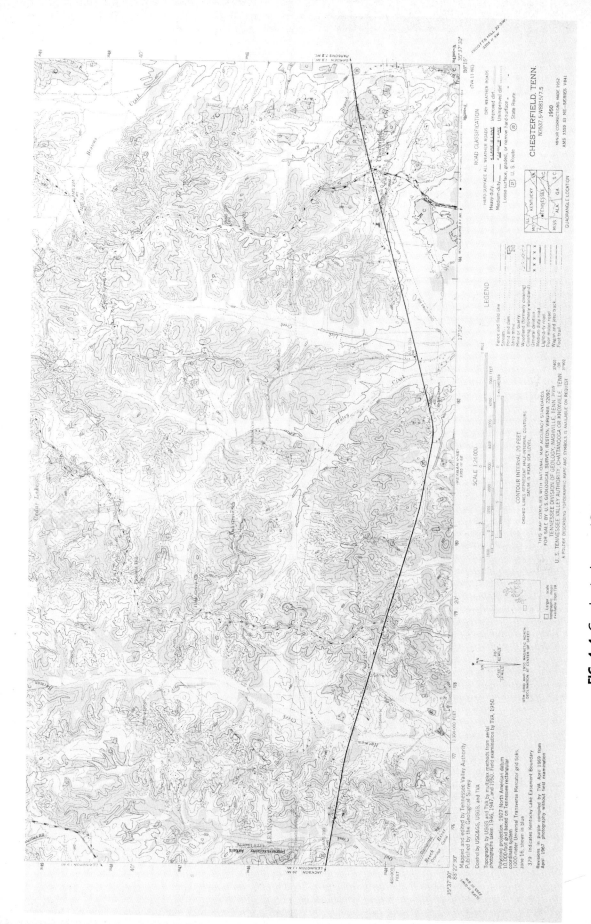

FIG. 1-4 Geological map (*Courtesy of Tennessee Valley Authority*)

surveys. More detailed and specialized maps, however, must be drawn from special surveys.

An example of how geodesic data are used to produce specialized maps is in the area of resource prospecting. In recent years, efforts have increased to identify and accurately locate sites of energy and mineral sources. Special survey projects and maps are prepared to determine the feasibility of prospecting and extracting mineral resources. Geodesic information is also helpful to locate mine workings and well drillings for the purpose of controlling and directing further prospecting and extraction activities.

1.2
TOPOGRAPHIC OCCUPATIONS

Many occupations are associated with the topographic and mapping field and in the development of related drawings. The production of topo drawings requires professional expertise ranging from collecting and analyzing source data to reproducing the topo itself. The occupational areas outlined in this section are those most frequently utilized in the production of topo drawings and maps. It should be noted that the job titles and descriptions listed here may be different from those used in some organizations. Job titles and descriptions are often tailored to meet the needs of the organization, based upon its size and the expertise of its technical staff.

Cartographer

The responsibilities assigned to *cartographers* range from supervisory functions to drawing and preparing maps. The assigned duties depend upon the formal training and practical experience of the individual. Cartographers are involved in the development and design of maps as well as the editing and monitoring of the final map product. They must be able to work cooperatively with other trained technicians such as cartographic and topographic drafters, photogrammetrists, mosaicists, and surveyors.

It is not uncommon to find supervisory-level cartographers who have advanced degrees in related disciplines. Hence, training varies from two-year, technical-school programs to advanced graduate programs. This occupation is considered to be at the professional-technical level and is typically filled by individuals who have both theoretical and hands-on competencies in topographic drawing. The Federal Office of Personnel Management lists two Government Service (GS) ratings for cartographers: GS-1371, for the two-year technician, and GS-1370, for the advanced, graduate school-prepared professional.

Photogrammetric Engineer

Photogrammetric engineering is a highly specialized occupation. The *photogrammetric engineer* is concerned with the planning, coordinating, and supervising of personnel who plan aerial surveys and direct photogrammetric compilations of topographic maps. Photogrammetric engineers must also be able to evaluate client needs, formulate mapping specifications, and prepare cost estimates. Expertise is essential in aerial photography, mission planning, photogrammetric instruments, and the application of remote sensing systems.

Photogrammetric engineers are also responsible for conducting and supervising research activities in surveying and mapping methods and procedures. Such activities are often accomplished through the application of photogrammetric map compilation, electronic data processing (EDP), and flight and control planning.

Individuals interested in entering this profession must have not only a strong engineering background, but also expertise in surveying techniques and photogrammetric practices.

Photogrammetrist

Photogrammetrists are technician-level personnel who have technical expertise in photogrammetric principles and procedures. Their duties can be broken down into three major areas. The first area is the analysis of source data related to the project. The source data frequently used consist of existing ground control, land surveys, and aerial photographs.

The second area of activity is the preparation and arrangement of maps, charts, and drawings from aerial photographs and other source data. This area includes the important phase of stereo compilation, involving the use of overlapping aerial photographs, taken from different angles, to create a three-dimensional image or model. The stereo model is used to clearly and accurately define various surface features.

Lastly, photogrammetrists are responsible for graphically identifying and noting all details shown on aerial photographs, such as control points, hydrography, topography, and cultural features.

Surveyor

Surveyors work with legal, engineering, architectural, and related constructing and mapping professionals to develop descriptions of tracts of land. Surveyors compile surface data such as angles, metes and bounds, points, and contours. They note and sketch all facts pertaining to the survey in a field manual. This manual is used in the development of topo drawings and specifications.

Surveyors are highly trained professionals who work with survey technicians and use specialized equipment to obtain accurate surface measurements.

Mosaicist

Mosaicists are responsible for examining aerial photographs and matching them into a photographic mosaic of a geographic area. To ensure accurate mosaics, these technicians calculate photographic scales and reference points that are used in matching one aerial photograph to another. Once a mosaic is put together, the mosaicist then refers to other photogrammetric products and surveying data to identify landmarks, topographic features, and existing structures.

Stereoptic Projection Topographer (Stereo Compiler)

Stereoptic projection topographers are also known as stereo-plotter operators, stereo operators, and stereo compilers. These technicians view aerial photographs stereoscopically in the production of topographic charts and maps. The *stereo plotter* is an instrument that produces two simultaneous projections of aerial photographs over the same geographic area, but from two different positions. Projecting the two photographs at the same time results in a three-dimensional model or image from which the operator delineates contours and other map features.

Cartographic and Topographic Drafters

The primary responsibility of cartographic and topographic drafters is the drawing of topo drawings, maps, and charts. *Cartographic drafters* have the major responsibility for drawing various types of maps. *Topographic drafters,* by comparison, limit their activities to preparing topographic maps. Both drafters frequently use and analyze survey data, source maps, and photographic details to determine the location and features of the geographic area. They are sometimes required to accompany survey teams into the field to compile data and identify the location of various landmarks. In some situations, the duties of these two types of drafters are the same.

1.3
THE WORKPLACE

Individuals trained to work in the mapping and topographic field find that there is a broad range of job opportunities and work environments available. Employment can be secured in both public and private sectors, such as state, federal, or local governments, and engineering firms. It has been found, however, that when trained in occupations such as topographic drafting, individuals are most likely to be hired by small organizations and have other duties assigned to them, such as mechanical drafting or serving as a member of a surveying crew.

Within this context, a *small organization* is defined as any firm or agency with 10 or fewer staff members, including technical, clerical, and support personnel. In small firms, the mapping and topographic specialist usually has a broad range of re-

sponsibilities and serves more as a generalist in a number of specialty areas.

People who want to specialize in a particular area of the mapping and topographic field should seek employment within a larger organization. Due to the nature of large organizations, it is possible to secure employment in specific positions with specific duties (for example, cartographer, topographic drafter, photogrammetrist, or stereo compiler). Large organizations tend to be more structured, and have specific lines of authority and responsibility.

Operating Procedures

Figures 1-5 and 1-6 are examples of organizational charts for a small organization and a large organization.

As can be seen, the small organization has fewer lines of communication and tends to be informal regarding office and field practices. An organizational standards manual is the exception rather than the rule in small organizations. Most information is passed along by word of mouth or by sim-

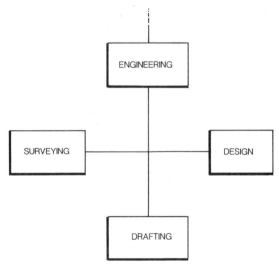

FIG. 1-5 Organizational chart of a small agency

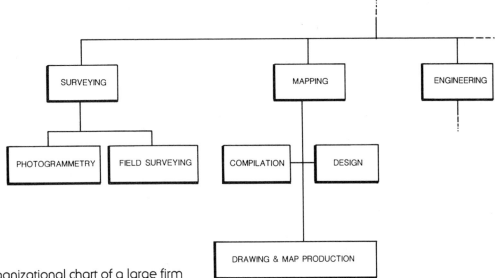

FIG. 1-6 Organizational chart of a large firm

ple memoranda. If there are any questions regarding responsibility, procedures, or technical information, staff members simply consult with the head engineer or cartographer, without worrying about following a chain of command.

Because of the nature and operation of the smaller organization, new employees are expected to start assuming job responsibilities within a short period of time. In larger firms, the individual is allowed more time to adapt to the work environment and learn the operational guidelines of the organization.

The larger the organization, the greater the number of positions and levels the employee is required to work through. Non-experienced drafters, for example, might have to start out as tracers or work on survey crews as rodmen before they are permitted to compile maps from source data. In addition, employees must work through specific lines of authority and communication. Organizational standards manuals are the rule rather than the exception in large organizations. These manuals include not only personnel policies, but also technical procedures and policies. Changes in policies or procedures are normally communicated through formal memoranda sent to all employees.

Self-employed Topographic Specialists

Individuals who are self-employed in the mapping and topographic field are usually experienced in their technical specialty and have work experience in the public and/or private sector. These professionals must have a high degree of skill and technical knowledge that is in demand by other firms or agencies.

Self-employed specialists are hired in situations where a firm does not have staff available or capable of handling a particular job. To be successful in self-employment, a person should be known in the field as a competent and productive professional and should have a proven track record. Because of the cycles involved in gaining contracts (peak and low times), self-employed topographic specialists frequently expand their area of work to include related areas such as architecture, engineering, surveying, and photography.

Topographic drawing is a procedure used to record the features of the surface of the earth. The topo drawer makes use of both mechanical and freehand drawing techniques. Topographic maps are used in numerous fields, such as surveying, engineering, hydrography, and geodesy.

Many occupations make use of and help in the preparation of topographic drawings. Cartographers help prepare and produce maps. Photogrammetric engineers supervise aerial surveys. Photogrammetrists analyze aerial photographs and help in the compilation of topographic maps, and surveyors provide surface data. Mosaicists compile series of aerial photographs for a photographic mosaic of a geographic region. Stereo compilers put together maps from three-dimensional stereo models. Cartographic and topographic drafters prepare the topo drawings.

Individuals trained in topographic and mapping skills can find employment in either the public or private sectors. Most, however, are employed in small firms and agencies. Those employed in large organizations tend to become more specialized in their duties and responsibilities. Those employed in small firms and agencies tend to become generalists, since their lines of authority and responsibility are informal and overlapping. Occasionally topographic and mapping specialists become self-employed if they so desire. To succeed, however, they must have a high degree of skills and knowledge, technical abilities, and a proven track record, in their field of specialty.

KEY TERMS

Aerial Photograph
Cartographer
Cartographic Drafter
Chart
Geodesy
Geological Map
Hydrographic Chart

Map
Mosaicist
Photogrammetric Engineer
Photogrammetrist
Source Data
Standards Manual
Stereo Compiler

Stereo Plotter
Survey
Surveyor
Topo
Topographic Drafter
Topographic Drawing
Topographic Map

1. Explain what is meant by the term *topographic drawing*.
2. Explain the meaning of the following phrase: "Topo drawing is neither an exact science nor a pure art form."
3. What are the applications of surveying in the development of topographic drawings?
4. Explain what hydrographic charts are, and their uses.
5. How are topographic drawings used in the field of engineering? List some examples.
6. Define *geodesy* and discuss its relationship to the mapping and topographic field.
7. Explain the primary duties involved in the following occupations:

   a. Cartographer
   b. Surveyor
   c. Stereo compiler
   d. Photogrammetrist
   e. Mosaicist
   f. Topographic Drafter

8. Identify the major differences between large and small firms; particularly discuss those differences in organizational structure and job responsibilities.
9. Describe the qualifications that are essential for someone who wants to become a self-employed map or topographic specialist.
10. Identify the major differences in the operating procedures between large and small organizations.

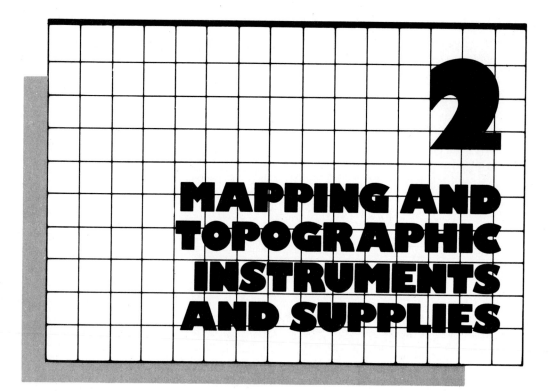

MAPPING AND TOPOGRAPHIC INSTRUMENTS AND SUPPLIES

The accuracy and type of information presented in topographic drawings affect the quality and financial practicality of many planning, engineering, and mapping projects. For these reasons, great care is taken in the design and preparation of topo drawings. In both the mapping and engineering fields, one way of assuring high-quality drawings is to use the best drawing and surveying instruments and supplies available, within practical limits.

The use of poor-quality or inappropriate instruments and supplies can cause the production of inaccurate and poor-quality drawings. Poorly drawn topos, in turn, detract from a product that is well conceived and designed. They also can steer a project toward financial ruin. Therefore, map drawers and drafters take care to select the tools and materials that are best suited for the task at hand.

There are many manufacturers who make and sell drawing and surveying instruments and materials. This creates a problem for the novice topographer: What equipment and supplies should be purchased, and from whom? To answer these questions, the inexperienced topographer should contact other professionals in the field. They can advise as to acceptable quality levels and can recommend supply companies that are able to meet the requirements.

There is a wide range of opinion as to what individual drawers and drafters should buy, and, in fact, if they should purchase their own equipment. In most firms, drafters are required to purchase their own set of basic drawing instruments and supplies. The employer, however, provides all required specialized and expensive equipment. Before drafters purchase an entire array of drawing instruments and materials, it would be to their advantage to first learn what their employer has and will supply. When you must purchase an item, the best rule to follow is to buy the best quality that you can afford, and maintain the item in good working order.

2.1
DRAWING MATERIALS AND MEDIA

A vast array of drawing materials and media is available on today's market. Some materials are considered to be essential in the drawing of any map or topo. Others can be used to improve the appearance and clarity of a drawing. The correct selection of materials and media is not only essential to ensure a quality product, but can also reduce drawing time and extend the life of the drawing itself. Seven types of drawing materials and media are discussed in this section. These are papers, cloth, film, leads and substitutes, inking equipment, and transfer overlays.

Papers

Papers, like cloth and film, are available in a variety of standard sizes of individual sheets or rolls, Table 2-1. Papers are used in a majority of cases to prepare maps, charts, and other types of topographic drawings. The papers used most commonly in the field are called *vellum*. They are also referred to as *tracing vellums* and *tracing*

papers. Vellum is used because of its ability to make high-quality drawings and good reproductive properties. In practice, there is little difference between vellum and tracing paper, since each is available in varying degrees of transparency and strength. Normally, the more transparent the paper, the higher the probability that it will be used as a tracing paper rather than as a medium for a master drawing.

Traditional tracing papers have characteristics based upon the weight, manufacturing process, and material. Tracing papers that are made of rag, or which have a high rag content, are fairly strong and permanent, but less transparent. On the other hand, if natural tracing paper is made of other cellulose fibers, it might be more transparent, but it will lack strength for drawings requiring permanency.

Vellums, or prepared tracing papers, were developed to combine transparency with strength. Vellums tend to be thicker than tracing papers, but they still possess the same degree of transparency. The transparency is made possible by the introduction of a synthetic resin. The transparency, durability, and relative inexpensiveness of vellums have made them a popular drawing medium among many drafters and cartographers.

Cloth

Cloth was the first standardized drawing medium. The original applications of cloth were in the drawing of maps and plans in ancient times. Cloth offers the combined advantages of transparency, surface quality, strength, and permanence. This medium can maintain surface quality regardless of the number of erasures and changes made. Furthermore, cloths are affected only to a small degree by

Table 2-1
Standard Sizes of Drawing Papers

Sheets:	Size in Inches	Size Designation
	8.5 × 11	A
	11 × 17	B
	12 × 18	—
	17 × 22	C
	18 × 24	—
	22 × 34	D
	24 × 36	—
	34 × 44	E

Rolls:	Width in Inches	Length in Yards
	8.5	20 and 50
	11	20 and 50
	18	20 and 50
	24	20 and 50
	30	20 and 50
	34	20 and 50
	36	20 and 50
	42	20 and 50
	54	20 and 50

age, and show no adverse effects in reproduction processes.

The original production of cloth required the use of high-quality linen and starch sizing, creating a high-quality surface capable of receiving either ink or pencil. Today's production techniques are different, however, due to the high cost of linen and the water solubility of starch. In place of linen, cotton is used, and a synthetic moisture-resistant sizing is used in place of starch.

Film

The most recent type of drawing medium used in the mapping and topographic field is *stable-base film*. Its increased use is due to its exceptional qualities of dimensional stability, strength, high transparency, and age and heat resistance. It is nonsoluble, waterproof, and has a high-quality appearance.

There are three major categories of films used in the production of maps and topos. These are polyester drawing film, unsensitized film, and sensitized film.

Polyester drawing film provides an excellent surface for accepting either ink or lead (pencil). Both produce deep, dark, and rich lines. When a pencil is used, it is recommended that the drawer move up to a harder lead (two or three degrees) than is used on paper. This is a common practice, because leads tend to wear much faster on polyester film than on paper or cloth. The only major disadvantage of polyester film is its relatively high cost.

Unsensitized film is used when stripping or peeling techniques are required. The use of a strippable or peelable film makes it possible to produce large areas of close-tolerance dimensions in a short period of time, such as in the case of drawing in a lake on a map. Traditionally, the outline of the lake would be drawn and the interior area inked in by hand. By using un-sensitized film, the outline of the lake would be cut and the interior area pulled off the backing. The film would then be placed on the drawn outline area of the lake, automatically adhering. This would leave a dimensionally exact area for either a positive or negative image.

Sensitized films are used in the reproduction process of maps and topos. They can be used to produce either negative or positive images. In addition to making reproductions of aerial photographs or drawings, many of the sensitized films that are available can be scribed. Once a map is reproduced on a sensitized sheet of film, which has a scribe coating, additional mapping data (such as contours, property lines, structures, and utilities) can be scribed directly onto the film.

Leads and Substitutes

The selection of an appropriate *lead* or *lead substitute* is critical in preparing high-quality drawings. As shown in Table 2-2, leads are available in a variety of hardness ratings. In total, there are 18 degrees of hardness that can be chosen for a given job.

The two softest leads are 6B and 5B. These are too soft for practical map and topo drawings, but they are excellent for some forms of freehand sketching. Leads between 4B and HB are useful for preparing certain types of renderings and sketches. Leads rated between F and 4H are normally used to prepare topographic drawings. Very hard leads, 5H and higher, are excellent for layout work and drawing reference lines that will not be reproduced.

Leads are used on vellum, cloth, and film. In situations where the drawing must be prepared on film, such as polyester, a special lead substitute can be used. The lead substitute is made of a synthetic base material unlike the graphite and clay mixture of leads. Synthetic lead is very adaptable to polyester film. It not only wears better but also eliminates the problem of lead smears on the drawing surface.

Lead holders are perhaps the most significant development in the use of leads in the drawing field. Pencils have been the traditional drawing tool of many map drawers and drafters. However, the *stick lead holder*, Figure 2-1, is rapidly replacing both the pencil and the mechanical holder. The most unique aspect of stick lead holders is their ability to draw lines at a consistent width without needing to be sharpened. The holders themselves are purchased according to lead width, which varies from

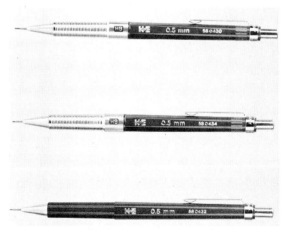

FIG. 2-1 Stick lead holders (*Courtesy of Keuffel & Esser Company*)

Table 2-2
Lead Hardness Ratings

Rating	Degree of Hardness
6B 5B 4B 3B 2B B	Soft
HB F H	Medium
2H 3H 4H	Hard
5H 6H 7H 8H 9H	Very Hard

0.3 millimeter (mm) to 0.9 mm. Each lead stick size is also available in each of the hardness ratings shown in Table 2-2.

Stick lead holders are operated by either pushing down or rotating the eraser end of the holder. The lead itself is stored in the barrel of the holder. It is automatically fed into the drawing head as needed. The consistency of the line width eliminates time-consuming sharpening processes and the associated mess.

Inking Equipment

There are times when a map drawing must be *camera ready*—of such quality that it can be photographed and reproduced. To ensure high-quality reproductions, drawings that must be camera ready are frequently prepared in ink. The advancements in inks and inking equipment have made this process much easier than it was in past years. In fact, there are some firms that routinely draw all their topos and plans in ink.

India ink was one of the first standard inks used in preparing camera-ready drawings. India ink, however, proved to be difficult to work with because it took a long time to dry. This caused numerous accidental smears and tended to clog up the nibs or points of the drawing instruments and pen. These problems have been solved with the availability of numerous types of inks. These inks dry quickly, yet are able to maintain their consistency in drawing instruments and pens so as to avoid clogging. Some inks are water soluble, while others are not. Many are available in a wide variety of colors.

Many inking instruments are available on today's market. One type is the *inking pen.*

Inking pens, like stick leads, come in a variety of standard widths, Table 2-3. As shown in Figures 2-2A and 2-2B, inking pens are available as technical fountain pens or as technical drawing pens. Fountain pens are used for artline and lettering work, while technical drawing pens are designed for freehand and mechanical line drawing.

Table 2-3
Inking Pens and Their Sizes

Type of Inking Pen	Available Line Widths	
Technical Fountain Pens	in Millimeters	
	0.1 0.2 0.3 0.4 0.5 0.6 0.8 1.0 1.25	
	Standard Sizes in Inches	**Metric Sizes in Millimeters**
Technical Drawing Pens	0.008 0.010 0.013 0.017 0.021 0.026 0.035 0.043 0.055 0.067 0.083 0.098 0.125 0.150 0.200 0.250	0.13 0.18 0.25 0.35 0.50 0.70 1.00 1.40 2.00

Mechanical instruments and lettering sets are also available for inking operations. Mechanical instruments such as compasses (Figure 2-3) are available with either a ruling pen or a technical pen adaptor. Lettering sets, Figure 2-4, are also available with inking adaptors. Lettering sets can be purchased in a variety of lettering styles and sizes.

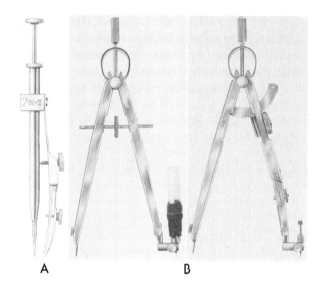

A   B

FIG. 2-3 Inking compasses. (A) Bow compass for drawing small-diameter circles, and (B) standard compasses with two types of inking adaptors (*Courtesy of Keuffel & Esser Company*)

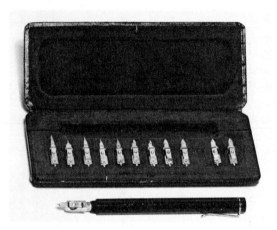

(A) Technical fountain pen set (*Courtesy of Keuffel & Esser Company*)

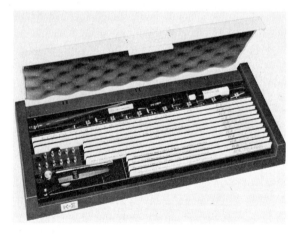

FIG. 2-4 Lettering set with inking attachments (*Courtesy of Keuffel & Esser Company*)

(B) Technical drawing pen set (*Courtesy of Keuffel & Esser Company*)
FIG. 2-2 Technical pen sets

Transfer Overlays

Transfer overlays are used to reduce the amount of time spent in preparing topo and map drawings while maintaining a high level of product quality. Transfer overlays were developed to provide a practical method for placing on drawings standardized symbols, expressions, or letters and numerals, Figure 2-5. Overlay symbols and letters can easily be transferred from a master sheet to the drawing. The overlay is placed on the proper location. Then the entire symbol is rubbed over with the use of a smooth stylist or blunt pencil point. This process transfers the symbol or letter to the paper and results in an excellent reproduceable copy. Over a period of time, however, transfer overlays tend to become brittle, and they do not have the same permanency as ink.

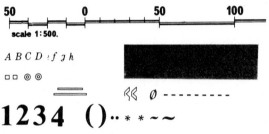

FIG. 2-5 Common transfer overlays

Specialty symbols, such as legends and logos, can be printed by supply companies upon special order. Larger figures, such as title blocks and some mapping scales, are usually printed on polyester film with a clear adhesive on the back. These overlays are then peeled off the master sheet and adhered to the drawing surface.

2.2 DRAWING INSTRUMENTS AND EQUIPMENT

Various types of drawing instruments and equipment are available to the map and topo drawer. Some are general in nature and are used in other fields such as mechanical and architectural drafting. Other instruments and equipment are designed specifically for mapping and topographic drawing. It is the purpose of this section to describe the basic drawing tools that are typically used in the mapping and topographic field.

Drawing Boards and Tables

Drawing boards and tables are available in different sizes, materials, and construction designs. A selection of the type of board or table should be based upon the type of work, drawing, and materials (such as vellum, film, or cloth) to be used on the drawing surface. Since topographic drawings can range in size from 8 inches by 10 inches (8″ × 10″) to 42″ × 84″ or larger, the size of either the board or table should be determined by the largest size of drawings that are commonly prepared.

Drawing boards are made of basswood or white pine, Figure 2-6. They are recommended for use on smaller drawings of 24″ × 36″ or less. Since drawing boards can easily be moved from one location to another, they are used in the field as portable equipment. Their portability makes it convenient to prepare preliminary drawings at a site location.

When portability is not a factor, *drawing tables* should be used, Figure 2-7. Unlike drawing boards, tables are considered to be stationary drafting room furniture. Being stationary, drawing tables have four basic advantages over drawing boards. First, the table surface is specifically constructed and made of materials that can hold up to a variety of drawing and related operations, such as scribing. Second, the drawing surface can be made extremely stable and fixed to ensure accuracy of the drawing and the transferring of measurements. Third, the table can be adjusted to various heights and drawing surface angles to suit

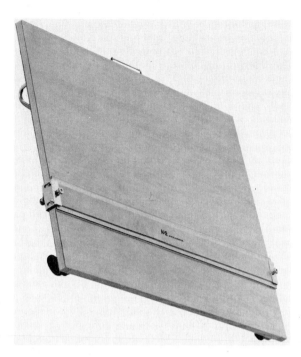

FIG. 2-6 Portable drawing board *(Courtesy of Keuffel & Esser Company)*

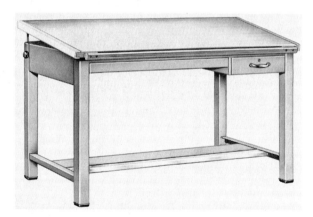

FIG. 2-7 Stationary drawing table *(Courtesy of Keuffel & Esser Company)*

the needs and comfort of the individual drawer. Finally, drawing tables can handle larger drawings with ease; the drawing tables are available in common surface sizes ranging from 32" × 42" to 43.5" × 84".

The *tracing table* is another type of drawing table used in mapping and topographic drafting. Tracing tables consist of a glass drawing top with a light source directly under it. When the light is on, the drafter is able to trace existing drawings easily. Tracing tables are also used extensively in scribing operations where transparent film is used, and when aerial photographs are used as a source of base data. The negative or positive film image is placed on the tracing table for easy interpretation or overlay drawing and scribing.

T-squares, Parallels, and Drafting Machines

The *T-square* is perhaps the most recognizable tool of the drafter. It is seldom used by topographic and map drafters, however. Some drafters may prefer the T-square to parallels and drafting machines, but most limit its use to preparing preliminary drawings at site locations.

Parallels are frequently used as guides for drawing horizontal lines, Figure 2-8. Available in standard sizes ranging from 42" to 96", they are constructed of either hardwood or metal with clear plastic drawing edges. Parallels are designed to move vertically up and down the drawing surface. They are kept parallel by wire guides attached to the corners of the drawing table.

Another unique feature of parallels is that they can be adjusted so as to not come in contact with the drawing surface. This is accomplished by attaching small rollers on the ends of the blade. The rollers raise the blade above the surface. This is an essential feature in cases where drafters are inking or scribing on coated film that could be easily scratched or damaged by dirt or dust collecting under the blade of the parallel.

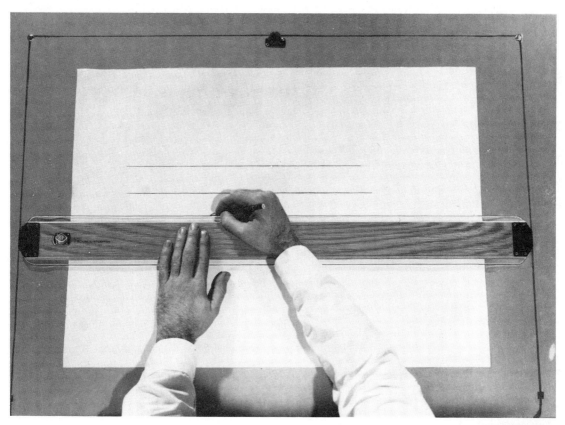

FIG. 2-8 Parallel mounted on a drawing table
(Courtesy of Keuffel & Esser Company)

Drafting machines are more flexible and adaptable to map and topographic drawing than are T-squares and parallels. Drafting machines can be used for drawing horizontal, vertical, and angular lines. This flexibility, therefore, provides the drawer with many time-saving options in both general and precision drawing.

There are two basic types of drafting machines: the standard and track types. The *standard drafting machine,* Figure 2-9, is the more common of the two within the general drafting field, where most drawings are in the 24″ × 36″ to 36″ × 48″ range. In the mapping and topographic field, however, many drawings exceed these dimensions. The *track-type drafting machine,* Figure 2-10, was designed to easily meet the requirements of both large and small sizes of drawings. Therefore, for many map and topo drawings, the track-type drafting machine might prove to be

FIG. 2-9 Standard drafting machine *(Courtesy of Keuffel & Esser Company)*

dering drafting machines, the type of scale desired (*i.e.*, civil, architectural, or metric) can be specified for each straightedge. This eliminates the frequent use of scales. Straightedges are available in common lengths varying from 12" to 36". Longer straightedges can be purchased by special order. Each drafting machine is also equipped with a rotating head and protractor, which can set the straightedges at increments of 0.5 degree (0.5°) when equipped with a double vernier. See Figure 2-11.

FIG. 2-10 Track-type drafting machine *(Courtesy of Keuffel & Esser Company)*

more advantageous than the standard drafting machine.

In both the standard and track-type drafting machines, the straightedges are made of either transparent plastic or aluminum. In a few cases, boxwood straightedges are available. These are preferred. When or-

Triangles and Templates

Triangles are used as straightedges in topographic and map drawing to draw vertical and standard angular lines, usually at 15° increments. There are three basic types of triangles: the 45°, 30°-60°, and adjustable triangles (see Figure 2-12). These triangles are all available in standard heights of 2" increments, ranging from 4" to 18". Typically, drafters purchase the 45° and 30°-60° triangles as a set. In such cases, it is good practice to make sure that the 30°-60° triangle is 2" longer than the 45° triangle. The adjustable triangle combines the functions of the triangle and protractor. In combination with one of the bases of the triangle, the hypotenuse can be set at any desired angle to within 0.5°. Because of this flexibility, some drafters prefer to use the adjustable triangle exclusively.

The three types of triangles are most frequently made of plastic, but are also available in wood and aluminum.

Templates are convenient, time-saving drafting tools used to draw a wide range of standard symbols and shapes. Some of the common templates used in the topographic and mapping field are general mapping symbols, civil engineer's radius

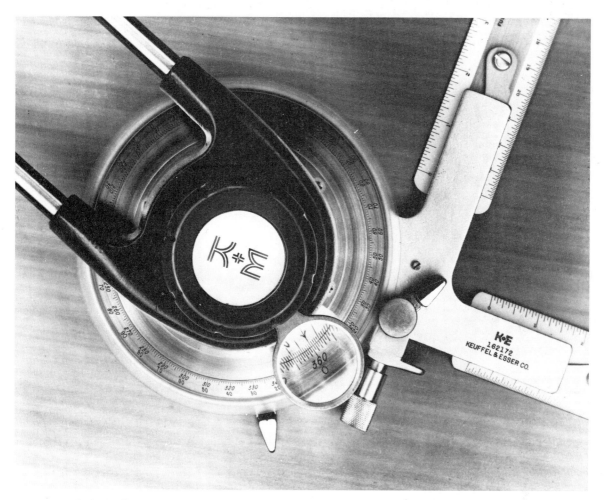

FIG. 2-11 Rotating head with double vernier *(Courtesy of Keuffel & Esser Company)*

guide, highway route symbols, and highway radius guide. Other templates common to all drafting and technical drawing fields are circle, ellipse, square, and lettering templates. Figure 2-13 shows some commonly used templates.

Guides and Curves

Guides are exceptionally useful to drafters for drawing lettering guide lines. The adjustable lettering guide, Figure 2-14, is specifically designed to draw guide lines of varying heights, including standard heights of ⅛", ¼", and ½". The side of the guide is angled at 68° for lettering slant letters.

A variety of curves can be found in drawing supply stores, but not all have application to topo drawing. Four common curves used in preparing topos are irregular, radius, railroad, and flexible curves. Examples of these are shown in Figure 2-15. *Irregular curves* are best suited for drawing curved lines that have variable slopes, such as contour lines. Irregular curves can be purchased individually or in a set. *Radius curves* are true curves with parallel edges. They allow for easy drawing of railroad or

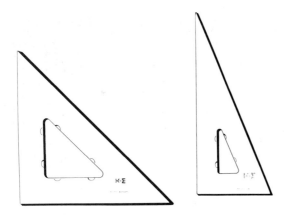

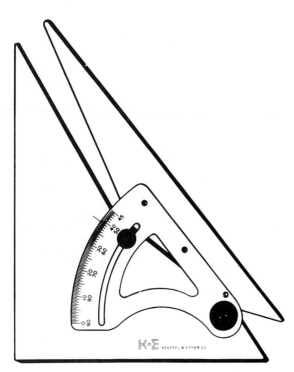

of a set of 55 curves marked to a scale of 1″ = 100′ − 0″. (1 inch = 100 feet) *Flexible curves* come in standard lengths ranging from 12″ to 30″, and can be adjusted to any curved slope. These curves lie flat on the drawing surface and are bent to fit the desired contour. Most flexible curves can be bent to fit any contour to within a 3″ radius or less.

Scales

The *scale* is one of the most important tools used in the drawing of maps and topos. It is essential, therefore, that professionals responsible for preparing these drawings become familiar and proficient in the use of scales. There are three types of scales available: the architect's, engineering, and metric scales. Of the three, the engineering and metric scales are more commonly used in the mapping and topographic field.

Engineering scales are also known as engineers' chain scales. These are perhaps the most frequently used scales in the mapping and topographic field. The engineering scale employs the use of a base 10 system of measurement. In other words, the inch is divided into divisions of 10, 20, 30, 40, and 50. This is unlike traditional scales, which are designed on a base system of 8 divisions per inch. The base 10 system is advantageous because most mapping measures are given in decimal and whole numbers, rather than in fractions and whole numbers. Engineering scales are available in 6″ and 12″ lengths. See Figure 2-16.

Metric scales are commonly found where maps and topos are prepared for international use, and are required in all European companies that produce maps and topos. The metric scale is not a standard found throughout the field in the United

FIG. 2-12 The 45°, 30°-60°, and adjustable triangles (*Courtesy of Keuffel & Esser Company*)

highway curves from a straight line to a tangency point. *Railroad curves* are similar to radius curves in that the edges are parallel to each other. These curves, however, were designed specifically for railroad layout work, and are composed

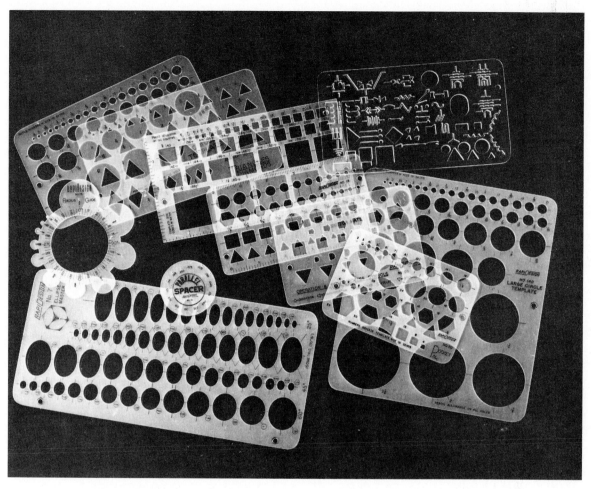

FIG. 2-13 Common templates *(Courtesy of Keuffel & Esser Company)*

States, but its use is increasing. The metric scale is similar to the engineering scale in that its entire measurement system is based upon 10 divisions per unit of measure. The scales, or graduations, found on the metric scale are based upon a ratio system of divided millimeters. The ratios commonly found on metric scales are 2 to 1 (2:1), 1:1, 1:2, 1:5, and 1:10. See Figure 2-17.

Architect's scales are primarily used in the architectural field for the preparation of building working drawings. At times, however, the scale is used to prepare plot and site plans, including contours and elevations, where foot and inch dimensions

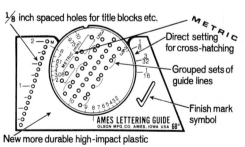

⅛ inch spaced holes for title blocks etc.

METRIC

Direct setting for cross-hatching

Grouped sets of guide lines

Finish mark symbol

AMES LETTERING GUIDE
OLSON MFG. CO. AMES, IOWA USA 68°

New more durable high-impact plastic

FIG. 2-14 Adjustable lettering guide *(Courtesy of Keuffel & Esser Company)*

may prove advantageous. The architect's scale is based on the English measurement system (*i.e.*, 12 inches to the foot, and

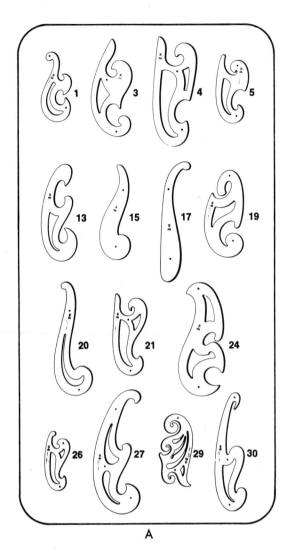

A

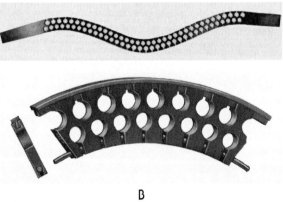

B

FIG. 2-15 Examples of two common curves. (A) Irregular, and (B) flexible *(Courtesy of Keuffel & Esser Company)*

8 divisions per inch). Architect's scales are available with 8 different scales. These are: $\frac{1}{8}'' = 1'$, $\frac{1}{4}'' = 1'$, $\frac{3}{8}'' = 1'$, $\frac{1}{2}'' = 1'$, $\frac{3}{4}'' = 1'$, $1'' = 1'$, $1\frac{1}{2}'' = 1'$, and $3'' = 1'$. See Figure 2-18.

Case Instruments

Case, or mechanical, *instruments*, Figure 2-19, should be selected on the basis of job requirements. Many of the drawing requirements in the mapping and topographic field largely depend on freehand drawing and the use of curves. Therefore, the use of mechanical instruments should be kept to a minimum. Case instruments are available with a variety of combinations of instruments, ranging from a simple set of 3 instruments to more than 15 pieces per set. Some drawers prefer to purchase each instrument on an individual basis, rather than as a set. In this way they can choose the best instrument available without being dependent upon a single manufacturing company. See Figure 2-19.

Scribing/Graving Instruments

The development and evolution of film and photographic materials created a need for specialized tools in the mapping and topographic drawing field. Specialized tools have been designed to work on both sensitized and unsensitized films. "Drawing" on sensitized film (for example, photographic negatives) is accomplished through scribers or gravers that are used to remove the sensitized coating. (This type of scribing is typically done on a scribe coat, where a light-sensitive coating is wiped on; a film negative is used to produce a guide image using high ultraviolet light. The image is then removed by scribing through the

FIG. 2-16 Engineering scale *(Courtesy of Keuffel & Esser Company)*

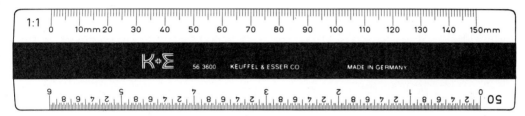

FIG. 2-17 Metric scale *(Courtesy of Keuffel & Esser Company)*

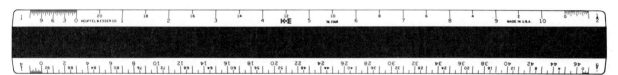

FIG. 2-18 Architect's scale *(Courtesy of Keuffel & Esser Company)*

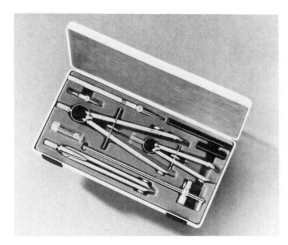

FIG. 2-19 Case instruments *(Courtesy of Keuffel & Esser Company)*

coating.) In comparison, when working with unsensitized film, cutting tools are used to actually cut through the film so that it can be lifted and removed.

Scribing/graving and cutting instruments are similar to other mechanical instruments, except that the drawing points are cutters or metal tips. Different types of tips and cutters are available for use on film media. The two most common types are the single- and double-line blades.

As shown in Figure 2-20, the basic mapping instrument set consists of relatively few pieces of equipment. These are the swivel graver, rigid graver, scriber holder, and scribers. The *swivel graver,* with an optic attachment, is used for graving topographic features and other information on negatives. The *rigid graver* is also used in the negative scribing process, when heavy or multiple lines are not required. The *scriber holder* is used as a pencil; it holds the tips or cutters. *Scribers* are the cutting or drawing points. They are used to scribe lines such as grids and contours.

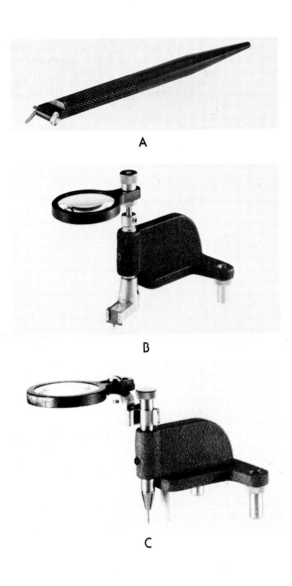

A

B

C

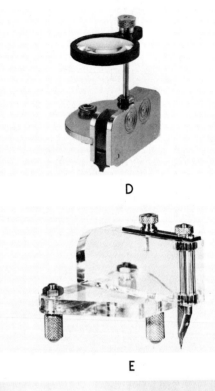

D

E

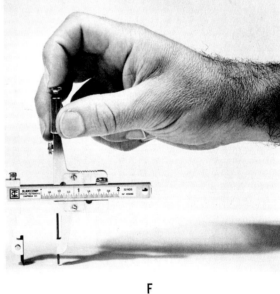

F

FIG. 2-20 Scribing/graving instruments. (A) Pen holder scriber, (B) swivel graver with optic, (C) rigid graver with optic, (D) straight line graver with optic, (E) dual rigid or swivel scriber, and (F) direct computing compass (*Courtesy of Keuffel & Esser Company*)

Planimeters

Planimeters, Figures 2-21A and 2-21B, are precision instruments used to measure plane areas of any shape. They operate by moving a tracer pin or lens around the boundary or circumference of the area. The area of the shape is then read directly from a measuring wheel or dial. The use of planimeters cuts down the amount of time

FIG. 2-21A Polar planimeter *(Courtesy of Keuffel & Esser Company)*

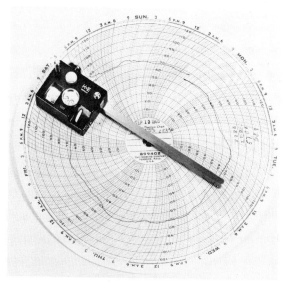

FIG. 2-21B Disc planimeter *(Courtesy of Keuffel & Esser Company)*

compensating polar planimeter and the *disc planimeter.* The compensating polar planimeter comes with fixed or adjustable arms. Fixed arm planimeters do not need setting adjustments, but they can only measure areas in square inches (in^2) or square centimeters (cm^2) and are then multiplied by a factor. The factor is determined by the scale of the map or drawing on which the shape is drawn or photographed. By comparison, the adjustable arm planimeter can be set to a variety of ratios so that the area can be read directly from the dial, in either the English or metric system, without the need for a factor multiplier.

The disc planimeter is considered to be a more precise instrument than the compensating polar planimeter. On the disc planimeter, a tracing wheel is used. It revolves approximately 10 times more per area of measurement than does the compensating polar planimeter. Hence, the disc planimeter is highly recommended for measuring area on maps or charts that have been reduced to a small scale.

2.3 SURVEYING INSTRUMENTS

The usefulness and applicability of maps and topographic drawings are dependent upon the accuracy of the source data used in their preparation. One method used to collect source data is the field survey. Surveys are the results of systematic procedures used to measure the relative features and characteristics of the earth's surface. The accuracy of surveys is dependent upon two critical factors: the quality and precision of the surveying instruments, and the competence of the surveying crew.

A great variety of surveying instruments is available on the market today. All types of

required to calculate the area of a shape, while increasing the accuracy. This method eliminates the need for counting squares or random dotted overlays in the calculation of plane areas.

Several types of planimeters are available which are applicable to mapping and topographic calculations. These are the

surveying instruments, however, can be categorized under one of two classifications. One classification is mechanical/optical distance (MOD) measuring instruments; the second classification is electronic distance measuring (EDM) instruments. The primary difference between MOD and EDM instruments is in the operation and information processing used by each, but MOD and EDM instruments have the same basic purpose and function.

This section presents descriptions of the basic surveying instruments used in the mapping and topographic field. The descriptions and operating principles discussed will help the reader to develop a comprehension of the differences between MOD and EDM instruments.

Transits

Transits, Figure 2-22, are instruments which receive their name from their ability to rotate up and down along the vertical plane (altitude) as well as the horizontal plane (azimuth). The process of rotating end for end in a vertical plane is known as *transitting*. A theodolite so designed is called a transit theodolite, or transit. The term *transit theodolite* is more common in European markets, while the term *transit* is usually used in the United States.

Transits consist of three major parts. The first is a sighting device known as an alidade. The alidade is usually equipped with vertical plane calibrations. The second part is an outer horizontal circle. The circle is a protractor held in the horizontal position and used to measure angles along the horizontal plane. The third part is the level. The level, or bubble level, is used to set the transit to a true horizontal plane.

It should be noted that alidades are sometimes used independently, apart from the transit. Plane table alidades (Figure 2-23) are used alone for mapping small areas and completing details between survey control points. Most plane

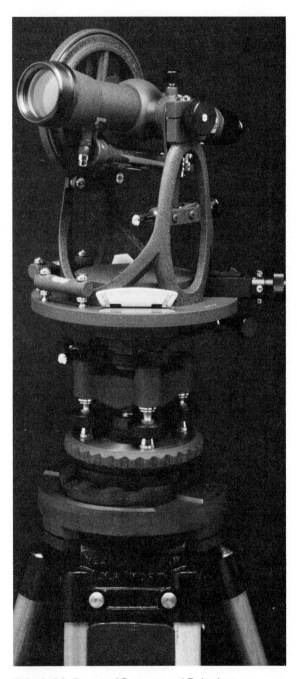

FIG. 2-22 Transit *(Courtesy of Teledyne Gurley)*

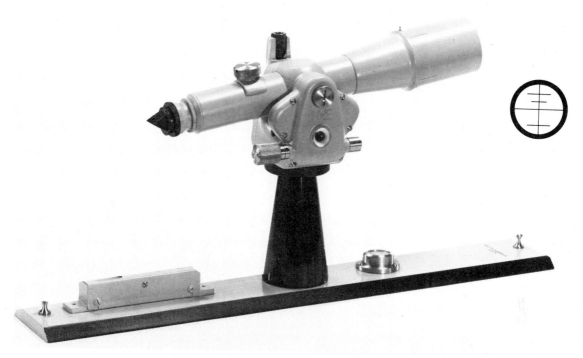

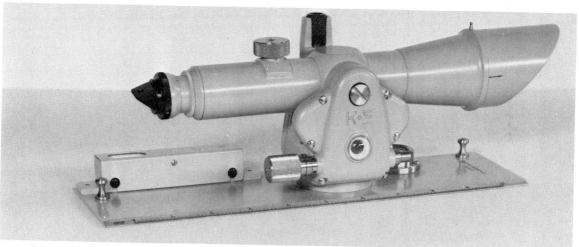

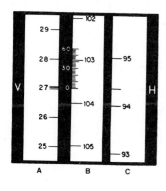

**View through
Scale-Reading Eyepiece**

A. Vertical Scale
reads 27.0

B. Elevation Angle Scale
reads 103° 42′

C. Horizontal Multiplier
reads 94.4

FIG. 2-23 Two plane table alidades, and scale *(Courtesy of Keuffel & Esser Company)*

table alidades are equipped with three scales. Viewed through the eyepiece, the scales are used to measure angles along the altitude and azimuth, and to determine horizontal multipliers.

Theodolites

In some cases it is hard to tell the difference between a transit and a theodolite. The *theodolite* is generally considered to be a highly refined optical reading transit. Theodolites are equipped with micrometer microscopes and/or electronic displays for reading angles with great precision. See Figure 2-24.

The basic operating principle of the theodolite is quite simple. Equipped with a horizontal circle similar to those used in transits, a pointer is used to indicate angle measurements. The pointer is set and pivoted at the center of the circle, so when the theodolite is turned and set at its target, the pointer will rotate along with it.

The size of a theodolite is determined by the diameter of the horizontal circle. For example, a 4″ theodolite has a horizontal circle with a 4″ diameter. A theodolite is also equipped with a level to set it to a true horizontal plane. The leveling process consists of using a level bubble in a glass tube shaped in the arc of a circle. When the theodolite is near the horizontal, with the concave side of the tube downward, the bubble of air will float to the top of the arc and center of the circle.

The need for more sophisticated engineering and measuring processes resulted in modifications and unique applications of theodolites. Scientists and engineers have used specialty and modified theodolites for aligning inertial guidance equipment, checking radar azimuths and elevation readouts, and calibrating dividing and rotating machines.

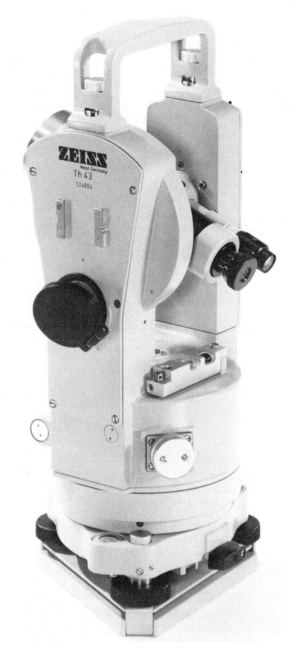

FIG. 2-24 Theodolite *(Courtesy of Zeiss)*

Range Finders

Range finders, are used to measure the bearing and distance of an object without having to go out to that object. This is par-

FIG. 2-25 Level *(Courtesy of Zeiss)*

for each classification are as follows: Short range finders measure distances up to approximately 100 meters (about 110 yards). Medium range finders measure distances from 50 meters to 1000 meters (approximately 55 yards to 0.62 mile). Long range finders measure distances from 250 meters to more than 10 kilometers (approximately 270 yards to more than 6.2 miles).

Levels

Levels are used to obtain the direct measurement of height differences between two points, Figure 2-25. Measurements are made by sighting a horizontal line of sight on graduated level rods that are held vertically over various points on a plat of land.

A level basically consists of a telescope similar to that found on theodolites. The telescope can be rotated along the horizontal plane, but not along the vertical plane. Thus, levels cannot be used to take altitude measurements directly. Attached to the level is a *bubble level,* which enables the line of sight to be brought into true horizontal.

Other Field Instruments

There are other field instruments that have been used for many years and which are still part of the surveying process. These instruments are tripods, plane tables, level rods, chains, and tapes. They are not as sophisticated as some of the other survey instruments described in this section, but they are important to collecting accurate source data.

Tripods, Figure 2-26, are the support structures upon which surveying instruments are mounted. The three legs of the tripod are made of either wood or metal. The

ticularly advantageous in situations where objects are either difficult or impossible to reach. The range finder does not measure angles of elevation; therefore, it cannot measure heights of objects.

Range finders used in surveying fit into three classifications: short, medium, and long range. As their name indicates, each classification is determined by the distances that can be measured. General guidelines

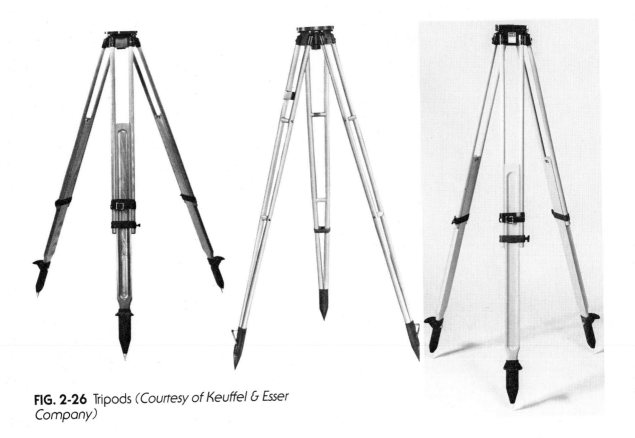

FIG. 2-26 Tripods *(Courtesy of Keuffel & Esser Company)*

legs are independently adjustable to provide a stable support on uneven terrain. *Plane tables* are small, flat surfaces that are mounted on tripods or other support structures. They are used only for topographic surveying. The plane table provides a flat surface to which surveying measurement instruments can be secured, permitting accurate measurements and plottings.

Level rods are constructed of two sliding wood sections. Each section consists of 0.01 graduations up to 13'. A target is also provided so that it can easily be spotted through a telescopic instrument.

Direct linear measures can be made by the use of chains or steel tapes. *Chains* are frequently used when surveys are being conducted for engineering projects. Chains come in lengths of 66' to 100', and are made of 100 pieces of straight wire joined together through a series of oval rings. The measures are taken from the brass handle ends along the chain. *Steel tapes,* Figure 2-27, are used more frequently than chains, and come in various lengths up to 100'.

MOD and EDM Instruments

Surveying instruments such as transits, theodolites, and range finders are available as either MOD or EDM instruments. Both MOD and EDM instruments are designed to accomplish the same type of surveying operation, but they use different techniques. MOD instruments use mechanical/optical principles, while EDM instruments use electronic principles.

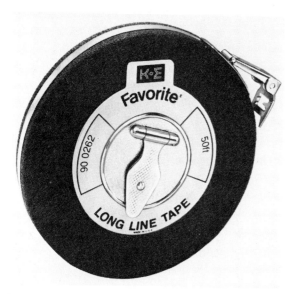

FIG. 2-27 Steel tape (*Courtesy of Keuffel &
Esser Company*)

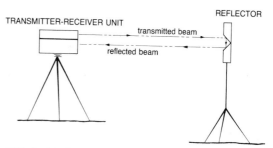

FIG. 2-29 Basic operation of EDM system

The typical MOD measuring system is
based upon instrumentation and procedures that can be traced back to ancient
civilizations. As illustrated in Figure 2-28,
MOD measuring systems require visual
sightings and the determination of distances and angles by optical and/or physical measurements. An example of this is
the determination of the distance and ele-
ation difference between two control
points, A and B. A transit would be set up
directly over point A, while a level rod
would be placed at point B. The distance

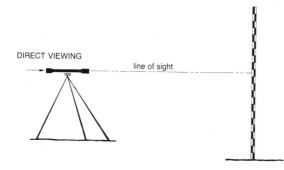

FIG. 2-28 Basic operation of MOD measuring
system

FIG. 2-30 Transmitter-receiver unit. This highly
sophisticated equipment allows for fast measurement, direct data processing in the field,
and automatic data flow. (*Courtesy of Zeiss*)

Table 2-4
Transmission Signals in EDM Systems

Transmission System	Transmission Method
Radio Wave	Pulse Type Continuous Microwave Modulation Cesium Standard (Atomic Clock) Antennas

Light Wave	Lasers
	  Ruby (Solid State)
	  He-Ne
	Luminescent Diodes
	Light Modulation
	Light Detection

between A and B would be physically measured by a chain or tape, while elevation difference would be sighted through the transit to the level rod.

The explosion of solid-state technology and microcomputers has made possible the rapid development and use of EDM instruments. Unlike MOD measuring instruments, EDM does not require the physical process of measuring linear distances, angles, or elevations. EDM relies upon the electrical transmission of signals, in the form of radio and/or light waves (see Table 2-4), and the translation of these transmissions into measurements.

Figure 2-29 is an illustration of how EDM systems operate. As can be observed, EDM systems are made up of three components—a *transmitter,* a *reflector* (sometimes referred to as a *retro-reflector*), and a *receiver.* To explain how an EDM system operates, we shall consider the same control points A and B described in the discussion of the MOD system. The dis-

tance and elevation would be determined in the following way using an EDM system. The transmitter and receiver, Figure 2-30, would be set up at point A, while the reflector would be located over point B. Next, the transmitter would send a transmission wave to the reflector, which would return it to the receiver. Inside the transmitter and receiver unit is a microcomputer that translates the time difference between wave transmission and reception into distance, and translates wave deflection into elevation.

Since an EDM system does not require the physical process of measuring, it decreases the number of members required on a survey crew. In fact, some EDM systems can be handled by one person. Because of this and the increase in speed and accuracy of measurements, EDM systems are rapidly replacing MOD measuring systems. Furthermore, as shown in Table 2-5, EDM systems have a wider range of application in surveying.

Table 2-5
MOD and EDM Systems Applications

System	Applications
MOD	Land to Land Surveying Land to Sea Surveying
EDM	Land to Land Surveying Land to Sea Surveying Land to Air Surveying Land to Satellite Surveying

2.4 SUMMARY

The production of high-quality and accurate topos is dependent upon the use of quality instruments and supplies in both the drawing and surveying processes. Many different kinds of drawing materials and media are used in the mapping and topographic field. These are papers, cloth, film, leads and lead substitutes, inking equipment, and transfer overlays.

The variety of drawing instruments is also great. Some drawing instruments are used in other areas of drafting and drawing, while some are specifically designed for the mapping and topo field. The basic drawing tools used in preparing maps and topos are drawing boards and tables, T-squares, parallels, and drafting machines. Others are triangles and templates, scales, case instruments, scribing/graving instruments, and planimeters.

Care should be taken when purchasing or selecting any instruments or materials. Furthermore, it is advisable for the beginning drawer to receive the recommendations of experienced professionals before purchasing equipment.

Surveying instruments are also applicable to map and topo preparation. Since the surveying process provides the source data from which all drawings are prepared, it is essential to use the most accurate piece of equipment available and practical. The basic surveying instruments used in the field are transits, theodolites, range finders, levels, tripods, level rods, chains, and tapes. All surveying instruments can be categorized as either mechanical/optical distance (MOD) measuring instruments or electronic distance measuring (EDM) instruments. The basic difference between the two is the way the measurement is accomplished.

KEY TERMS

Architect's Scale
Camera Ready
Case Instrument
Chain
Cloth
Curve
Drawing Board
Drawing Table
EDM
Engineering Scale
Fountain Pen
India Ink
Lead
Lead Holder

Lead Substitute
Level
Level Rod
Lettering Guide
Metric Scale
MOD
Parallel
Plane Table
Planimeter
Polyester Film
Range Finder
Scribing/Graving Instrument
Sensitized Film
Standard Drafting Machine

Stick Lead Holder
Tape
Technical Drawing Pen
Template
Theodolite
Tracing Paper
Track-type Drafting Machine
Transfer Overlay
Transit
Triangle
Tripod
T-square
Unsensitized Film
Vellum

REVIEW

1.   Discuss the factors to be considered when deciding whether or not to purchase drawing or surveying instruments.

2.   Explain the difference between vellum and tracing paper.

3.   What was drawing cloth originally made of, and what is drawing cloth made of today?

4.   Name the advantages and disadvantages of using polyester film as a drawing medium.

5.   Explain the concepts of lead hardness and hardness rating.

6.   How do the inks used today differ from traditional India ink?

7.   What are transfer overlays, and how can they reduce drawing time?

8.   Under what condition might a drafter want to use a drawing board and T-square?

9.   What is the major function of parallels, and how do they work?

10.   What are the two types of drafting machines found in drawing rooms?

11.   Identify the three common types of triangles used in preparing drawings.

12.   Name two types of curves that are particularly suited for use in map and topo drawing.

13.   Name the three types of scales used in drafting. Which type is most often used in preparing map and topo drawings in the United States, and which is used in European companies?

14. What are case instruments?

15. Identify the drawing medium for which scribing/graving instruments are best suited.

16. Name two types of planimeters, and explain their uses.

17. What do *MOD* and *EDM* stand for, and what is the basic difference between the two?

18. Explain the difference between a transit and a transit theodolite.

19. Explain the purpose of each of the following surveying instruments:

      a. Transit   d. Theodolite

      b. Range finder   e. Level

      c. Plane table   f. Level rod

20. Briefly explain how an EDM system operates.

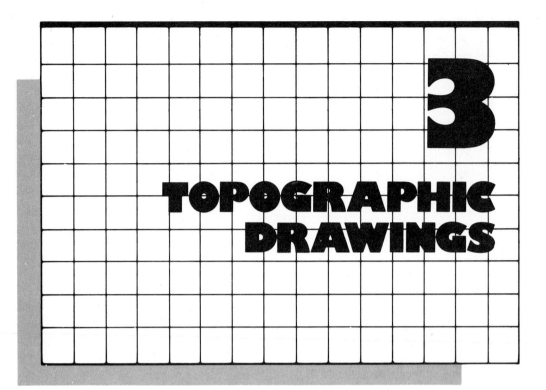

3

TOPOGRAPHIC DRAWINGS

Topographic drawings have different, though overlapping, connotations to geographers and engineers. In the geography field, topographic drawings are viewed as the traditional maps usually published by private and public agencies. Published maps are available as either individual map sheets, or as a collection of maps known as an *atlas*.

In contrast, topographic drawings prepared for the engineering and planning field are usually not published and sold. The topos are drawn by and for the individual firm or agency, and tend to be drawn at a larger scale. Topographic drawings used in engineering and planning are typically scaled at 1:50 to 1:400.

It is important to note that the use and meaning of symbols and scaling are consist-

ent in both the geography and engineering fields. However, the procedures used for data collection and reproduction techniques are different.

In this chapter we discuss the application of topographic drawings in the engineering and planning field.

3.1
THE TOPOGRAPHIC DRAWING

Topographic drawings are the graphical representations of data gathered by surveying techniques. Surveying procedures are used to find the exact location of natural and artificial characteristics of an area of land. These characteristics, or *intricacies*, are then represented as symbols and scaled onto a topographic drawing or map.

Natural characteristics are features that are found in nature, such as rivers, lakes, hills, and valleys. *Artificial* or *non-natural characteristics* are those features that have been constructed or modified by human hands or machines. Buildings, roads, canals, cemeteries, and utility systems are examples of artificial features.

The location of terrain characteristics is known as *planimetry.* This process should not be confused with a *planimeter,* which is an instrument used to measure plane areas. The representation of terrain, or the "lay of the land," is called *topography.*

A topographic drawing, then, is a drawing that shows the lay of the land by the use of scaled symbols.

3.2
SURVEYING AND TOPOGRAPHIC DRAWINGS

Surveying and topographic drawings are closely interrelated. All maps are plotted from data collected by various surveying techniques ranging from field surveying to photogrammetric data collection. The drawing is a visual method of presenting the surveying data. A visual presentation is much faster and easier to understand and interpret than a series of linear, area, volume, and angular measurements.

Surveying data provide information about the horizontal and vertical characteristics of land. Since both types of data are recorded, the topographic drawing must represent three-dimensional information on a two-dimensional plane (that is, the drawing surface). This is accomplished by using a combination of symbols and notations.

Topographic drawings show the elevations of individual features as well as the distances between them. All physical and cultural features are also shown, as identified by precise engineering surveys and field inspection. Topos, therefore, show the location and shape of mountains, valleys, plains, streams, lakes, and artificial objects.

In some situations it is too expensive for an individual firm or governmental organization (such as a small city or township) to survey a large area of land. To overcome this problem, various agencies provide survey data at a minimal cost. Table 3-1 lists some of these agencies and the types of surveying data that are available.

Another practice, which is gaining wide acceptance among smaller cities and townships, is to conduct a detailed survey of a jurisdiction and then sell the data to developers and construction firms. The price charged is minimal, and is based upon the land area to be developed. This approach has proved to be advantageous to both the city and the developer. Not only does the city provide a service to developers, but it can also keep track of what is being planned for development. Another advantage is that the developer does not have to pay for an expensive survey. This is especially critical for smaller areas of land.

The aerial photograph is the most common format of data sold.

3.3
THE USE OF CONTROL POINTS

The accuracy and usefulness of any topographic drawing or map rely upon the use of a fixed reference known as a *control point.* Control points are reference markers whose location and elevation are known. Increased accuracy of many maps is possible by utilizing a number of control points together. Therefore, most legal descriptions, such as deeds and filed site plans, are

Table 3-1
Sources of Surveying Data

Agency	Data Available
Army Map Service (Department of the Army)	Topographic maps and charts
Board of Engineers for Rivers and Harbors (Department of the Army)	Maps and charts of navigable rivers and harbors
Bureau of Land Management	Survey progress map of the United States, maps of public lands and reservations, township plots, and land divisions
Coastal Engineering Research Center (Department of the Army)	Coastal erosion data
Corps of Engineers (Department of the Army)	Topographic maps, charts of major water systems, and Great Lakes
International Boundary Commission, United States and Canada	Topographic maps for areas ½ mile to 2-½ miles on either side of the United States-Canadian border
Local Municipalities	Street maps, zoning maps, drainage and utility maps, and horizontal and vertical control data
Mississippi River Commission (Department of the Army)	Flood-control data, and various hydrographic and hydraulic studies
National Ocean Survey (Department of Commerce)	Topographic maps and coastal charts, hydrographic and topographic data of inland lakes and reservoirs, various control data, aeronautical charts, seismological and magnetic data, coast pilots' data, tide and current tables, bench mark locations, level data, and elevation tables

Table 3-1, *Continued*

Agency	Data Available
Naval Oceanographic Office (Department of the Navy)	Nautical charts, aeronautical charts, and manuals on navigation
Soil Conservation Service (Department of Agriculture)	Soil maps, charts, and indexes
United States Forest Service (Department of Agriculture)	Forest reserve maps and vegetation classification
United States Geological Survey (Department of the Interior)	Topographic maps and indexes, bench mark and level data, elevation tables, horizontal control and monument data, geologic maps, water resources and streamflow data
United States Postal Service	Rural free delivery (RFD) maps per county

based upon the use of several control points.

Permanent control points are also referred to as *monuments*. A monument can be either natural or artificial. Examples of natural monuments are rocks, trees, and geological formations. Artificial monuments are usually marked on the ground by metal tablets 2 inches to 4 inches in diameter. These tablets are set in rock or masonry, or on driven metal rods. Monuments are frequently indicated by symbol on topographic drawings.

Control surveys are surveys that use monuments or establish their own control points. Control surveys are conducted to describe the terrain and surface features in correct relationship to other surface features. During a control survey, two types of control measurements are made: *horizontal measurements* and *vertical measurements*.

Horizontal ground control is used to establish and maintain correct positioning, scaling, and orientation of features drawn on a map. Horizontal measures are made by using another form of fixed reference — latitude and longitude measurements. Latitude and longitude measurements are made for selected points within an area of land, and are determined and located by the field team. Latitudes and longitudes are used as a series of references for the topographic drafter.

Vertical ground control is used to determine the location and shape of contours. The elevations made in the field can be based upon one or two reference points. The first zero level that can be used is the median sea level (feet above or below sea level). The second zero level can be a given control point (feet above or below control point A). Two zero levels are used in some situations. This is discussed in further detail in Chapter 7.

Horizontal and vertical control points are the framework upon which map detailing is based. In addition to accurately locating surface features and objects, control points make it possible to key and cross-reference various drawings. Thus, by using a standard system of control referencing, map detailing can be accurately matched from one drawing to another.

3.4 DRAWING PROCEDURES

Topographic drawing and mapping procedures have changed considerably since the days when topographic maps were sketched by hand in the field. Today, most maps are compiled in an office by photogrammetric methods, involving aerial photographs and complex stereoscopic plotting instruments. In some situations, the drafter has been replaced by computer graphics systems.

Even with the advances made in photogrammetry and computer graphics, there is still a significant need for the field survey crew and topographic drafter. This is particularly true for smaller developments and tracts of land where detailed photogrammetric procedures and computer programming would prove to be too cumbersome and time consuming. Such a situation is the small-scale map typically produced in the field of geography. Small-scale maps are produced almost totally by the use of high-elevation aerial and satellite photographs. Hence, the "drawing" of mapping details is actually scribed onto photographic positives that are printed onto polyester film.

A situation that might require the additional use of field survey teams, for small-scale maps, is where the vegetation coverage would make it difficult to calculate contour intervals by photogrammetric procedures. A case in point is a photogrammetric mission performed over a tract of agricultural land that was to be converted into a highway. Due to the growth of crop vegetation, the photographic mission incorrectly calculated the amount of fill required by 2 feet (the height of the crop). To prevent this problem from occurring again, the engineering firm established the now standard practice of sending a field survey team over any area where vegetation growth could cause elevation calculation errors.

While it is true that the advancements in computerized graphics have displaced some drafters, the need for hand-drawn topographic drawings and maps is still significant. The use of computerized graphics is advantageous where there are relatively few readings. With an extensive number of readings, however, the time required to feed the data into a computer exceeds the time required to draw the topo or map by hand. Thus, we can see that the need for drafters will continue.

Topographic drawings and maps should give as complete a picture of the terrain as can be produced legibly. Standards for content have been developed to meet the needs of most mapping projects, and help to make the drawing legible. Drawing procedures are discussed more fully in later sections of this book. At this point, however, some basics of topographic drawing and map production are described.

Inking Drawings
Topographic drawings are drawn in pencil or ink, as well as by scribing. Pencil drawings utilize the same procedures and technicalities as other engineering drawings, while scribing is reserved primarily for published maps.

Ink drawings, however, require special consideration. Because of the significant number of ink drawings required, the following guidelines should be considered.

Naturally, thorough advance planning is basic to the production of a good ink drawing. The drafter must first plan the drawing, and then follow a time-efficient routine of inking procedures.

A basic requirement is to provide as much clarity and legibility as possible. This is particularly important when it is necessary to superimpose one type of information upon another.

Symbols indicating features such as towns, houses, and roads have little flexibility for location. Therefore, these symbols should be inked first.

The degree of importance of each topographic feature must be kept in mind constantly. The relative importance of symbols, for example, can be conveyed by increasing the thickness of the lines, thereby emphasizing the "blackness" of the symbol. Once the most significant elements have been drawn in, the drafter can determine the levels of shading for the rest of the drawing. This will identify the sequence of specifying information (drawing from darkest to lightest).

Lettering Names and Notes

The type of lettering used for topographic drawings should be conservative and non-ornamental. Most topos prepared for engineering and planning purposes are lettered by freehand methods. It is there-fore important for the drafter to develop a clear and legible lettering style. When projects are presented to the lay public or to financial backers, the drawings often have mechanical lettering or transfer lettering superimposed onto the drawing.

Hand lettering should reflect a distinctive technique, and the lettering should be in balance with the rest of the drawing. Distribution skills are extremely important to obtaining a balance in topographic drawing and map lettering. A few guidelines to follow when lettering are presented here.

Try to postition the lettering so that it will read from the bottom of the drawing, and use a lettering guide to establish the height and alignment of letters. Use vertical and inclined lettering in appropriate places. Be sure that the lettering spacing in words and throughout the drawing is balanced.

Show true or magnetic North by a north arrow, regardless of the drawing's size.

Names should designate the specific object, and must be clear and easy to read. Letter the names in parallel with the lower border line. The greater the number of names required, the greater the need for graduation in the lettering scale, from most prominent to least prominent.

A legend should be used to designate names that are too lengthy to fit attractively on the drawing.

Railroads, roads, waterways, and other types of communication systems should be parallel and close to the right-of-way boundary markings. For waterways, forests, or swamps with elongated outlines, try to center the name and extend it in the direction of the greatest elongation.

3.5 SUMMARY

Topographic drawings are prepared differently for their two primary users, geographers and engineers. Topographic drawings prepared for geographers are usually published, as individual maps or in a collection called an atlas. Drawings prepared for engineering and planning firms are usually not published, and are larger-scale drawings.

The data collected by surveying procedures are critical to the drawing of accurate maps. The location of terrain intricacies is known as planimetry, while the graphical representation of the terrain is called topography.

Surveying practices used by engineers provide both horizontal and vertical data to the drafter. Horizontal data give information about linear and area measurements. Vertical data provide information about elevation. When these are used in combination, angular and volume measures can be obtained.

The accuracy of topographic drawings is based upon the use of control points. Control points are reference markers whose location and elevations are known. The references can be either natural or artificial objects.

Topographic drawings can be produced by hand, by computer assistance, or by photogrammetric procedures. Though computerized graphics and photogrammetry are making significant strides in map production, drafters still produce many of the drawings required in engineering and planning firms.

Topographic drawings can be produced by pencil, ink, or scribing. Pencil is used most commonly for engineering projects, while scribing is used most often for published maps. Inking, however, is used when a special presentation is required.

KEY TERMS

Atlas	Control Survey	Planimetry
Control Measurement	Horizontal Measurement	Topography
Control Point	Monument	Vertical Measurement

REVIEW

1. What is the difference between a published map and an unpublished map? Explain why there is this difference.

2. What are the basic differences in the topographic drawings used in the geography field and the engineering field?

3. Explain why surveying is so critical to topographic drawing.

4. Describe control points, and explain how they can increase the accuracy of a topographic drawing.

5. What are the three drawing methods used for preparing topographic drawings? Identify which procedure is used most frequently for published maps, for engineering projects, and for presentations.

4

CONVENTIONS AND PRACTICES

Established conventions and practices are used in mapping and topographic drawing to convey information simply and clearly. The preparation of topographic drawings is an important aspect of cartographic and engineering services firms. Such organizations require personnel who have highly developed skills in graphics and graphic communications. To ensure that map and topo drawings can be understood and interpreted, a series of standardized symbols, abbreviations, and procedures have been developed.

Professional and technical personnel concerned with topographic data often rely heavily upon drawings and maps for the bulk of their information. Maps and drawings, therefore, must be expressive and adaptable to a variety of uses. In this chapter, therefore, we discuss the conventions and practices widely used in map and topo drawing.

4.1 SYMBOLS

In map and topographic drawing, a graphic system is used to present a variety of information. Along with obvious information (for example, cities, rivers, and elevations) presented as symbols, a great deal of "indirect" information is communicated. Indirect information is presented through the way the map or topo is designed and drawn. This includes the emphasis of particular types of data and the general arrangement of the graphic system.

Most of the information presented on maps and topographic drawings is in the form of symbols taken from a standardized system of graphic representations. As a result, any physical phenomenon (abstract or concrete) can be identified on a drawing. The graphic symbolism used to identify these phenomena can be categorized into four major groups on the basis of symbol form: point, line, area, and volume.

A *point symbol* is used to identify a location, and is theoretically nondimensional. A point symbol can be any of a number of graphical marks that identify a place or feature on the drawing, such as an elevation point, city, historical marker, or control point. As can be surmised, point symbols do occupy an actual geographic area and frequently cover considerable land area, but are only used to indicate the location of that area or specific point. Figure 4-1 shows examples of point symbols.

Line symbols are used to identify features that are elongated, narrow, and geometrically consistent. They are also typically used to identify phenomena that connect one point to another. Examples of features represented by line symbols are roads, rivers, railroad lines, and air routes. In comparison to point symbols, line symbols are considered to be theoretically one dimensional in character. Examples of line symbols are shown in Figure 4-2.

Area symbols are used to identify polygons (pieces of land); they are defined as being theoretically two dimensional. Area symbols help describe the characteristics of an area of land. Characteristics may include the political boundaries involved (national, state, county, municipal), the vegetation covering the land (woodlands, grasslands, various types of crops), or selected hydrographic features of an area (rivers, ponds, lakes). Because area symbols are two dimensional, they are more descriptive than point and line symbols, and they require more artistic and technical skill. Figure 4-3 shows examples of area symbols.

A *volume symbol* depicts spatial variation in the amount or quantity of a variable. Volume symbols are three-dimensional representations and can be used for a variety of data presentations.

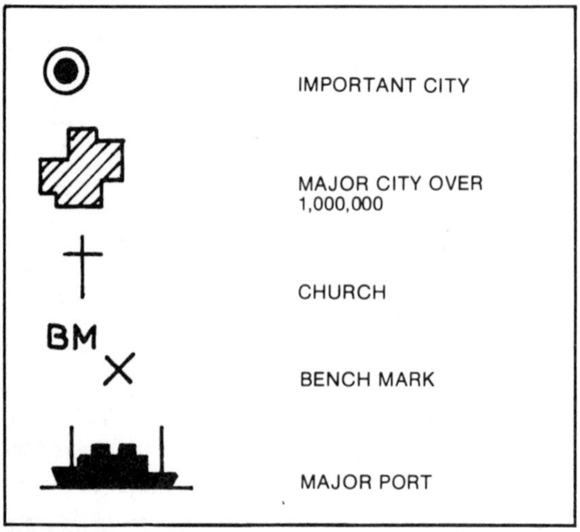

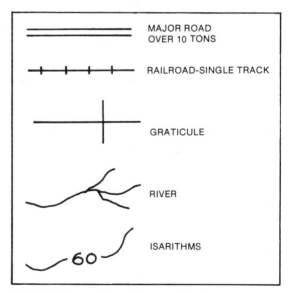

FIG. 4-1 Examples of point symbols

IMPORTANT CITY

MAJOR CITY OVER 1,000,000

CHURCH

BENCH MARK

MAJOR PORT

FIG. 4-2 Examples of line symbols

MAJOR ROAD OVER 10 TONS

RAILROAD-SINGLE TRACK

GRATICULE

RIVER

ISARITHMS

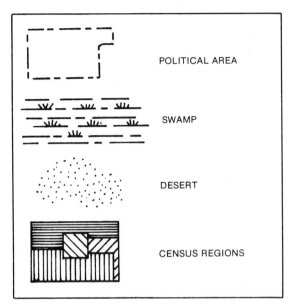

FIG. 4-3 Examples of area symbols

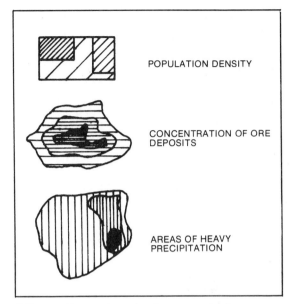

FIG. 4-4 Examples of volume symbols

Examples of volume symbols are a set of isolenes representing the amount of annual rainfall in a particular area, concentration or amount of vegetation or population, and the amount of land at a particular elevation. As can be seen in Figure 4-4, volume symbols require more drawing and technical skills than are required to draw the other types of symbols.

Drawing Considerations for Symbols

The use of symbols is limited in the drawing of topographic maps. The limitations are dictated by the size of the map and the amount of detail that is to be shown. Hence, the symbols used when designing and drawing topos should meet certain criteria. The symbols should

- be easy to read and understand.
- conform to the general design and purpose of the drawing.
- follow standard and acceptable formats.

In light of these three criteria, there are four factors that must be taken into account during map preparation and drawing. These factors are composition, proportion, shading, and color.

Composition pertains to how the map drawing is designed in relation to its proposed use. For example, if a map is to be used primarily for identifying contours and relief, all other symbols (such as hydrographic, vegetation, or boundaries) would be drawn in lightly. Contours, however, would be drawn in as dark lines. Furthermore, if one is using standardized symbols and follows a set of symbol specifications (line weights, etc.), the contrast relationships will be guided by these standards and will emphasize those features that are important to the user. For an example of composition, see Figure 4-5.

In addition to using symbol specifications, it is good practice for the beginning drafter to draw the most important map features *first*. Using our example in Figure 4-5, this means that all contour lines and relief

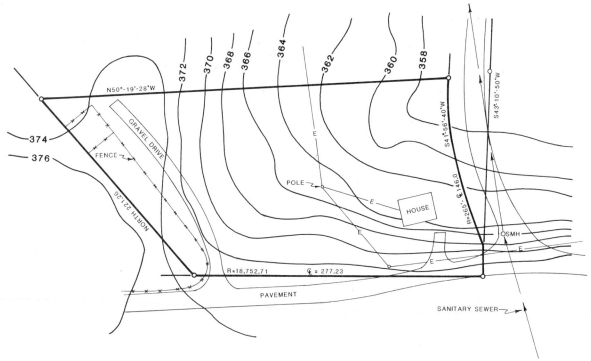

FIG. 4-5 Topo with primary emphasis on contours

symbols would be drawn first. The remaining map symbols would then be drawn in the order of their importance. By using this simple procedure, both novice and experienced drafters will provide a simple quality-control system to deemphasize or even eliminate unimportant information.

Proportion refers to drawing the size of the symbol in relation to the size of the area being described. If a symbol is drawn too large, for example, it may appear to be a point symbol rather than an area symbol. On the other hand, if a symbol is drawn too small it may be difficult to read and interpret. Figure 4-6 presents examples of correct and incorrect proportioning techniques.

Shading is a technique used to clarify and/or emphasize a particular symbol. Shading, when properly used, gives the symbol a perception of depth. Shading

should be drawn as if the sun's rays were coming down at a 45° angle. Unlike the other factors discussed in this section, shading should be limited to landscape and large-scale topographic drawings. See Figure 4-7 for an example of shading.

Color is of primary importance in the field of cartography, but is infrequently used in engineering and surveying. Its primary use in engineering drawings is in preparing presentations for prospective clients or financial backers. In cartography, however, it is frequently used as a method of differentiating between various map symbols and topographic features. Like all symbols, color topography should follow standard mapping conventions. Common colors used in mapping, and examples of their application, are presented in Tables 4-1 and 4-2.

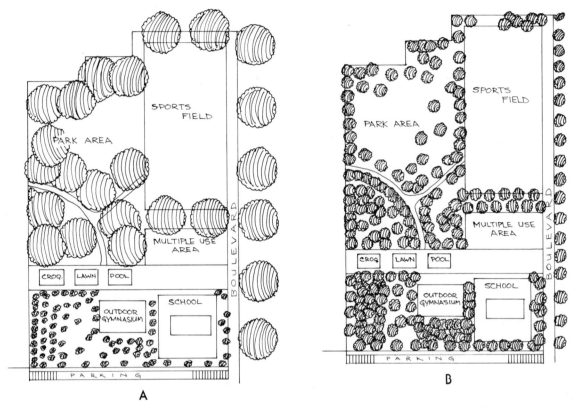

FIG. 4-6 Correct and incorrect proportioning techniques using symbols (**A**) The proportions used for the trees are too large and too small. (**B**) The correct proportioning presents a clearer and more accurate presentation of the site.

Topographic Map Symbols

Symbols, like topographic maps and drawings, are categorized according to characteristics and scaling. *Characteristics* are the types of information given (i.e., hydrographical or surface forms), while *scaling* can be either small-scale or large-scale maps. The topographic map symbols that are found on small-scale drawings are discussed in this section. This is not to say that these symbols are not used in large-scale drawings (or vice versa) but are more common on small-scale mapping.

Maps drawn for use in the United States must conform to standards established by the United States Geological Survey.

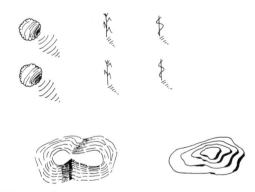

FIG. 4-7 Use of shading for symbols

These standards (National Map Accuracy Standards—NMAS) govern map accuracy as well as symbol specifications. They are

Table 4-1
Color Usage in Topographic Mapping

Color	Possible Tints	Examples of Use
Black		Boundaries Site Plan Lines Man-made Features Point Information Lettering and Notations
Blue	Light Dark	Seas Lakes Glaciers Rivers and Streams Canals Swamps Oceans Drainage Systems
Brown	Reddish Dark	Contours Relief and Hypsographic Features Rock Outcrops (sometimes) Additional Details
Green		Woodland Coverage Scrub Vineyards and Orchards General Vegetation
Yellow		Additional Boundaries Distributional Tones
Red	Dark Light	Important Roads Field Lines Public Land and Land Grants Subdivisions Urban Areas
Grey		Shading Hatch Marks
Purple		Office Revisions from Aerial Photographs

Table 4-2
Color Usage for Elevation Indicators

Color	Elevation Range in Feet
Dark Green	0 to 1000
Light Green	1000 to 2000
Pale Brown (Light Tan)	2000 to 3000
Light Brown	3000 to 5000
Medium Brown	5000 to 7000
Deep Brown	7000 to 9000
Dark Brown	More than 9000

the recognized authority upon which topographic maps and their symbols are drawn. It should be noted, however, that older maps and those produced in other countries may have some variations from the U. S. Geological Survey guidelines. Examples of these variations are shown in Appendix A.

For our discussion purposes only, all small-scale topographic map symbols are broken down into eight broad categories. The categories are (1) road transportation systems, (2) railroad and waterway transportation systems, (3) non-natural objects, (4) map references, (5) boundaries, (6) surface forms, (7) hydrographic forms, and (8) other surface characteristics. A majority of these symbols are produced in multicolor form.

Road transportation system symbols include representations for all forms of roads, streets, and trails. See Figure 4-8. These symbols include primary and secondary roads, light-duty and improved roads, unimproved roads, proposed roads and roads under construction, dual or divided roads/highways, and trails.

Road transportation symbols are all considered to be line symbols that connect one location (phenomenon) to another. On multicolor maps, major roads should be drawn in red; all other roads should be drawn in black. For single-color maps (black), major roads can be drawn in as shown in Figure 4-8.

Compilers, scribers, and drafters must take care to preserve alignment of roads and other linear features.

Railroad and waterway transportation system symbols include graphic symbols that represent railroad systems, waterway systems, and related waterway objects. See Figure 4-9. Examples of railroad system symbols are single and multiple tracks, juxtapositioned railroads, narrow gage tracks, and overpasses. Examples of waterway system symbols are canals and locks, dams, and waterways with roads.

Like road transportation symbols, railroad and waterway transportation system symbols are line symbols (except for the water area symbols). On all maps, multicolor and single color, these symbols are drawn in black lines. The only variance is the blue coloration for an area of water.

Non-natural object symbols pertain to the representation of any object that does not occur as a natural phenomenon. Non-natural objects are constructed objects, such as buildings, cemeteries, communication

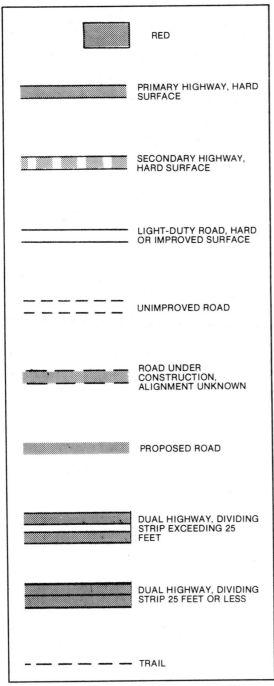

FIG. 4-8 Road transportation system symbols

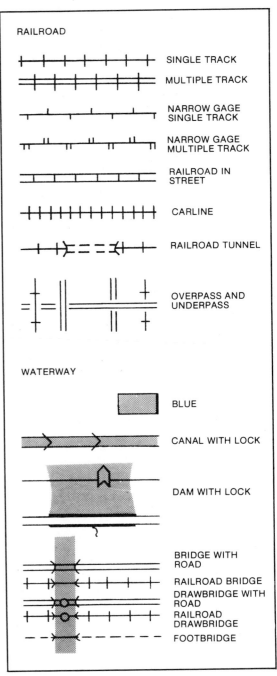

FIG. 4-9 Railroad and waterway transportation system symbols

lines, mining operations, and storage tanks. See Figure 4-10.

Non-natural object symbols can be either point, line, or area symbols. When drawing these symbols, regardless of the colorations used, only black lines should be used. Hence, even on multicolor maps these symbols appear in black.

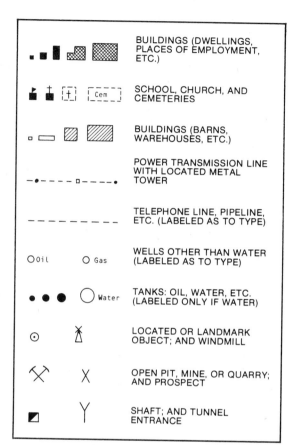

	BUILDINGS (DWELLINGS, PLACES OF EMPLOYMENT, ETC.)
	SCHOOL, CHURCH, AND CEMETERIES
	BUILDINGS (BARNS, WAREHOUSES, ETC.)
	POWER TRANSMISSION LINE WITH LOCATED METAL TOWER
	TELEPHONE LINE, PIPELINE, ETC. (LABELED AS TO TYPE)
Oil Gas	WELLS OTHER THAN WATER (LABELED AS TO TYPE)
Water	TANKS: OIL, WATER, ETC. (LABELED ONLY IF WATER)
	LOCATED OR LANDMARK OBJECT; AND WINDMILL
	OPEN PIT, MINE, OR QUARRY; AND PROSPECT
	SHAFT; AND TUNNEL ENTRANCE

FIG. 4-10 Non-natural object symbols

BM △ 6733	TABLET, SPIRIT LEVEL ELEVATION (HORIZONTAL & VERTICAL CONTROL STATION)
△ 4372	OTHER RECOVERABLE MARK, SPIRIT ELEVATION LEVEL (HORIZONTAL & VERTICAL CONTROL STATION)
VABM △ 1245	TABLET, VERTICAL ANGLE ELEVATION (HORIZONTAL CONTROL STATION)
△ 4123	ANY RECOVERABLE MARK, VERTICAL ANGLE OR CHECK ELEVATION (HORIZONTAL CONTROL STATION)
BM X 631	TABLET, SPIRIT LEVEL ELEVATION (VERTICAL CONTROL STATION)
X 361	OTHER RECOVERABLE MARK, SPIRIT LEVEL ELEVATION (VERTICAL CONTROL STATION)
X 638 X 638 (BROWN)	SPOT ELEVATION
750 750 (BLUE)	WATER ELEVATION

FIG. 4-11 Map reference symbols

Map reference symbols function as reference and data points for the reader. Examples of these symbols are horizontal and vertical control stations, elevation check points, and spot elevations. See Figure 4-11. Unlike other symbols, however, these symbols also provide alpha-numeric data.

Map reference symbols are considered to be point symbols. When drawn on multicolor maps, they can be shown as either black or colored markings. Control stations should always appear in black, spot (land) elevations in brown, and water elevations in blue.

Boundary symbols are used to identify political and nonpolitical delineations or boundaries. Examples of boundary symbols are national, state, county, section, and fence lines. See Figure 4-12.

Boundary symbols are considered to be area symbols since they define two-dimensional phenomena. In multicolor mapping, boundaries appear in either red or black, depending upon the type of boundary depicted. Major boundaries, such as state, county, and other governmental areas are always shown as black lines. Boundaries that are less important, such as survey township and section or range lines, and nongovernmental areas, are shown as red lines.

Surface form symbols provide information that describe the surface characteristics of land. Examples of these symbols are contours, fills, mine dumps, washes, tailings, sand, and gravel areas. See Figure 4-13.

All surface form symbols are either area or volume symbols. Contours are usually

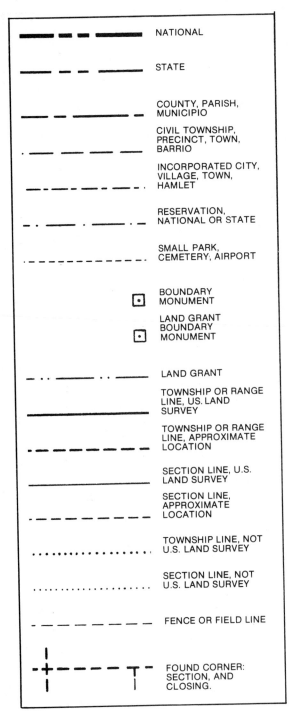

	NATIONAL
	STATE
	COUNTY, PARISH, MUNICIPIO
	CIVIL TOWNSHIP, PRECINCT, TOWN, BARRIO
	INCORPORATED CITY, VILLAGE, TOWN, HAMLET
	RESERVATION, NATIONAL OR STATE
	SMALL PARK, CEMETERY, AIRPORT
	BOUNDARY MONUMENT
	LAND GRANT BOUNDARY MONUMENT
	LAND GRANT
	TOWNSHIP OR RANGE LINE, U.S. LAND SURVEY
	TOWNSHIP OR RANGE LINE, APPROXIMATE LOCATION
	SECTION LINE, U.S. LAND SURVEY
	SECTION LINE, APPROXIMATE LOCATION
	TOWNSHIP LINE, NOT U.S. LAND SURVEY
	SECTION LINE, NOT U.S. LAND SURVEY
	FENCE OR FIELD LINE
	FOUND CORNER: SECTION, AND CLOSING.

FIG. 4-12 Boundary symbols

considered to be volume symbols, while all others are area symbols. When drawn on multicolor maps, these symbols should be

drawn in brown, since they represent land characteristics.

Hydrographic form symbols are used to identify and describe the characteristics of areas of water. These symbols can be used to identify either standing or flowing water areas. Examples of hydrographic symbols are rivers, streams, aqueducts, rapids, falls, lakes, and depth curves. See Figure 4-14.

Hydrographic symbols can be either line or area symbols in representation. In multi-color maps, hydrographic symbols are usually drawn in blue. There are only two exceptions to using blue throughout the symbol. The first is where constructed or other non-natural objects are located in the hydrographic area. The second is for dry lake beds—the outline is in blue, but the area is drawn in with the color brown.

Other surface characteristic symbols include any remaining topographic map symbols used to describe surface features. These symbols include non-natural and natural characteristics as well as land and water areas. Examples of these symbols are urban areas, vegetation characteristics, and marshes. See Figure 4-15.

Symbols in this category are all represented in nonblack colors on multicolor maps. Following consistent practice, those symbols representing water areas are in blue, vegetation areas are in green, vegetation in water areas are in blue and green, and urban areas are represented in red or in a red tint. When a map is prepared in black only, such features are drawn as shown in Figure 4-15 and are tied in with a map legend.

Detailed Topographic Symbols

Large-scale maps and drawings often require information that is more detailed than can be provided by using small-scale

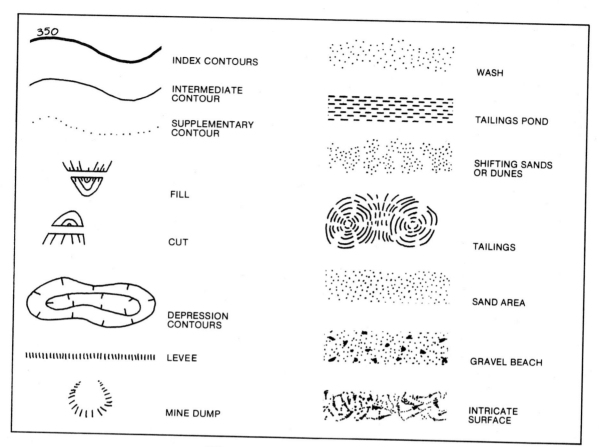

FIG. 4-13 Surface form symbols

topographic map symbols. Therefore, a series of symbols has been developed for use in large-scale drawings. Detailed topographic symbols are found most frequently in engineering topos and site plans, rather than in cartographic map drawings.

The wide range of detailed topographic symbols makes it necessary for us to divide them into eight broad categories. The categories, divided for our discussion purposes only, are land and relief forms, geological structures, hydrographic and navigation, vegetation, boundaries, civil structures, gas and oil works, and service systems.

Land and Relief Form Symbols. These are used to describe the terrain within an area of land. Examples of features that can be distinguished with these symbols are rock outcrops, cliffs, reefs, washes, levees, bluffs, and ledges, Figure 4-16.

Geological Structure Symbols. These are used to describe the geology of the land. In a way, geological structure symbols are a form of scientific notation that identify geologic structures. These symbols are divided into seven major groupings: altitude and direction, bedding, cleavage and schistosity, joints, linear elements, folds, and faults. See Figure 4-17.

Hydrographic and Navigation Symbols. These are frequently found on navigation and hydrographic charts for navigable waterways and recreational areas. Examples of these symbols are depth curves, nautical symbols, and navigational aids.

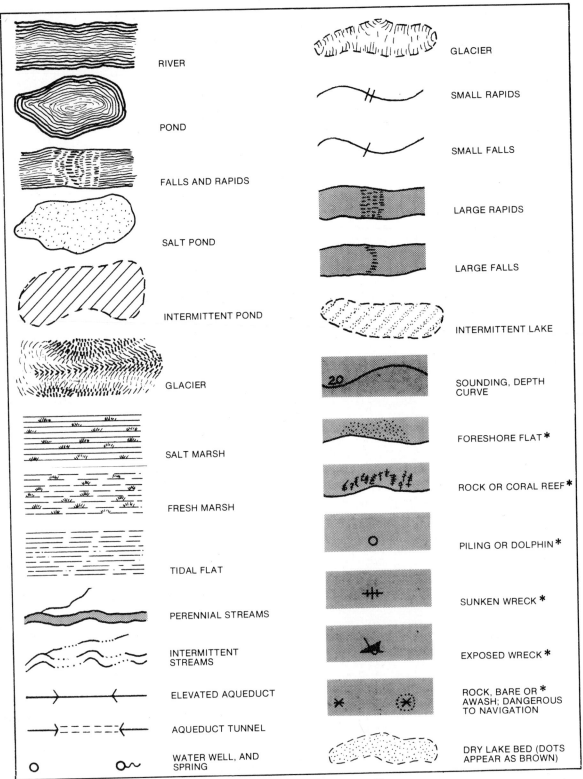

RIVER

POND

FALLS AND RAPIDS

SALT POND

INTERMITTENT POND

GLACIER

SALT MARSH

FRESH MARSH

TIDAL FLAT

PERENNIAL STREAMS

INTERMITTENT STREAMS

ELEVATED AQUEDUCT

AQUEDUCT TUNNEL

WATER WELL, AND SPRING

GLACIER

SMALL RAPIDS

SMALL FALLS

LARGE RAPIDS

LARGE FALLS

INTERMITTENT LAKE

SOUNDING, DEPTH CURVE

FORESHORE FLAT *

ROCK OR CORAL REEF *

PILING OR DOLPHIN *

SUNKEN WRECK *

EXPOSED WRECK *

ROCK, BARE OR * AWASH; DANGEROUS TO NAVIGATION

DRY LAKE BED (DOTS APPEAR AS BROWN)

*NOTE: All symbols inside shading appear as black. All other lines and shading appear as blue.

FIG. 4-14 Hydrographic form symbols

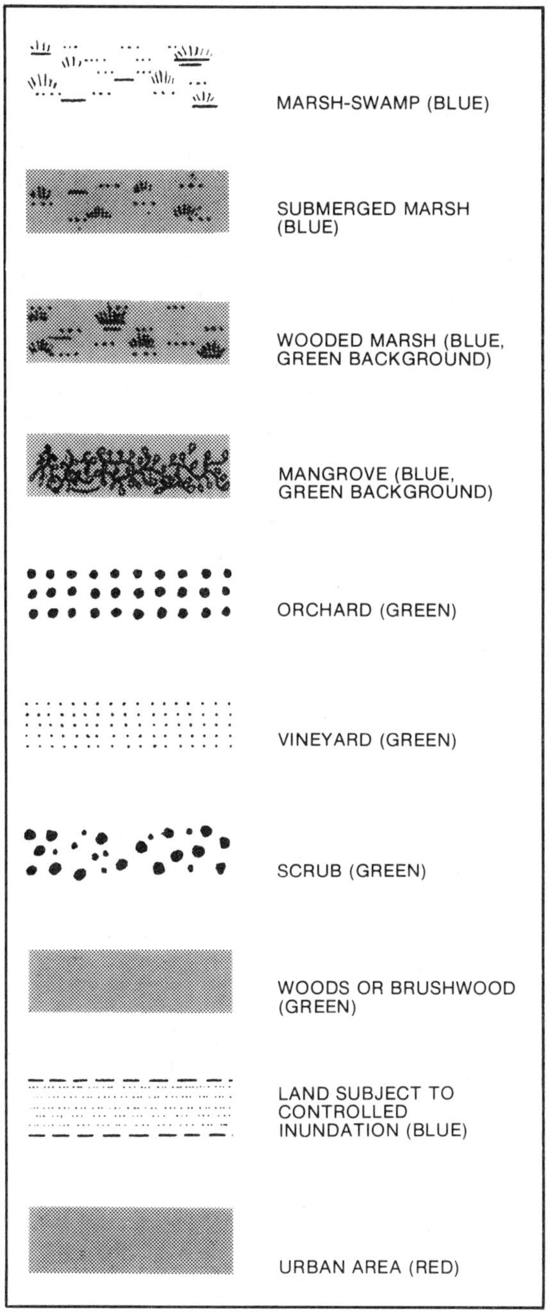

MARSH-SWAMP (BLUE)
SUBMERGED MARSH (BLUE)
WOODED MARSH (BLUE, GREEN BACKGROUND)
MANGROVE (BLUE, GREEN BACKGROUND)
ORCHARD (GREEN)
VINEYARD (GREEN)
SCRUB (GREEN)
WOODS OR BRUSHWOOD (GREEN)
LAND SUBJECT TO CONTROLLED INUNDATION (BLUE)
URBAN AREA (RED)

FIG. 4-15 Other surface characteristic symbols

The best sources for up-to-date information on hydrographic navigation charts and charting are the National Ocean Survey and Lake Survey Center. See Figure 4-18.

Vegetation Symbols. This category includes representations for plants that are found natural in nature as well as those planted and cultivated for personal or commercial use, Figure 4-19. Examples of these symbols are vineyards, orchards, grasses, trees, brush, corn, and cleared land.

Boundary Symbols. These include both legal and informal boundaries. Legal boundaries include city, township, and property lines. Informal boundaries are fences, hedges, and survey lines. See Figure 4-20.

Civil Structure Symbols. These symbols represent any object or structure that was planned and constructed, Figure 4-21. They do not represent any natural object. Examples of civil structure symbols are buildings, mines, tunnels, and oil wells.

Gas and Oil Works Symbols. Symbols are used to identify active and inactive facilities used in the gas and oil industry. Examples of these symbols are rigs, drilling wells, small oil wells, salt wells, dry holes, gas wells, and abandoned wells, Figure 4-22.

Service System Symbols. These are used to describe systems such as pipe lines, electrical systems, and culverts. Examples of these symbols are sewer lines, water tanks, water lines, gas pipes, steam pipes, oil pipes, valves, meters, feeder lines, and switches. See Figure 4-23.

4.2 ANNOTATIONS

Topographic maps and drawings often require more information than can be provided by using symbols only. Notations, names and titles, map scaling, measurements, and similar data must also be

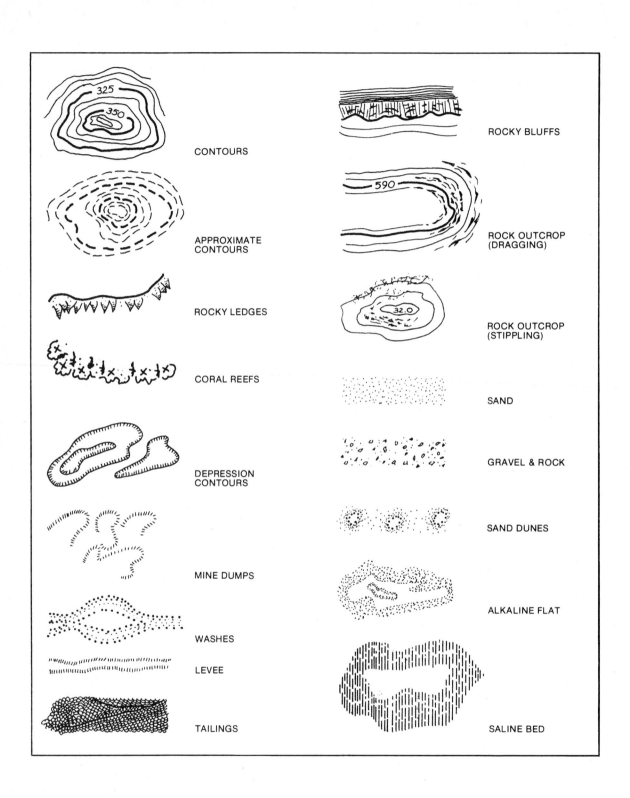

FIG. 4-16 Land and relief form symbols

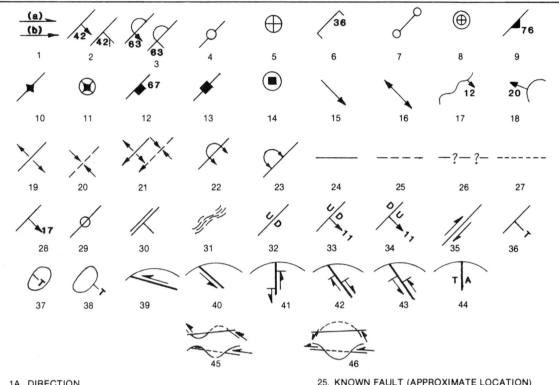

1A. DIRECTION
1B. ATTITUDE
2. STRIKE & DIP OF BEDS (ARROW USED IN DETAILED MAPS)
3. STRIKE & DIP OF OVERTURNED BEDS
4. STRIKE OF VERTICAL BEDS
5. HORIZONTAL BEDS
6. STRIKE & DIP OF CLEAVAGE (SLATE)
7. STRIKE OF VERTICAL CLEAVAGE (SLATE)
8. HORIZONTAL CLEAVAGE (SLATE)
9. STRIKE AND DIP OF SCHISTOSITY OR FOLIATION
10. STRIKE OF VERTICAL SCHISTOSITY OR FOLIATION
11. HORIZONTAL SCHISTOSITY
12. STRIKE & DIP OF JOINT PLANE
13. STRIKE OF VERTICAL JOINT PLANE
14. HORIZONTAL JOINT PLANE
15. DIRECTION OF PITCH OF LINEAR PARALLELISM, FLOW LINES, LINEAR STRETCHING, OR ALIGNMENT OF MINERALS & INCLUSIONS
16. DIRECTION OF HORIZONTAL LINEAR ELEMENT
17. GENERAL STRIKE & DIP OF MINUTELY FOLDED BEDS
18. DIRECTION OF PITCH OF MINOR FOLDS (NATURE OF ISOCLINAL FOLD AT ITS PLUNGING END)
19. AXIS OF ANTICLINE
20. AXIS OF SYNCLINE
21. PITCH OF AXIS OF ANTICLINE OR SYNCLINE
22. AXIS OF OVERTURNED OR RECUMBENT ANTICLINE (SHOWING DIRECTION OF INCLINATION OF AXIAL PLANE)
23. AXIS OF OVERTURNED OR RECUMBENT SYNCLINE (SHOWING DIRECTION OF INCLINATION OF AXIAL PLANE)
24. KNOWN FAULT

25. KNOWN FAULT (APPROXIMATE LOCATION)
26. DOUBTFUL OR HYPOTHETICAL FAULT
27. CONCEALED FAULT COVERED BY LATER DEPOSITS (KNOWN OR HYPOTHETICAL)
28. DIP OF FAULT PLANE
29. VERTICAL FAULT PLANE
30. SHEAR ZONE STRIKE & DIP
31. SHEAR ZONE
32. U-UPTHROW, HIGH-ANGLE FAULT . D- DOWNTHROW, HIGH ANGLE FAULT
33. NORMAL FAULT
34. REVERSE FAULT
35. DIRECTION OF HORIZONTAL MOVEMENT IN SHEAR FAULT, TEAR FAULT, OR FLAW
36. OVERTHRUST LOW-ANGLE FAULT. T, OVERTHRUST SIDE
37. KLIPPE OR OUTLIER REMNANT OF LOW ANGLE FAULT PLATE. T, OVERTHRUST SIDE
38. WINDOW, FENSTER, OR HOLE IN OVERTHRUST PLATE. T, OVERTHRUST SIDE.
39. OVERTHRUST; LOW-ANGLE FAULT; ARROW IS DIRECTIONAL MOVEMENT OF ACTIVE BLOCK
40. UNDERTHRUST; LOW-ANGLE FAULT; ARROW IS DIRECTIONAL MOVEMENT OF ACTIVE BLOCK
41. VERTICAL; HIGH ANGLE FAULT (ARROW SHOWS DIRECTIONAL MOVEMENT)
42. NORMAL FAULT; HIGH ANGLE FAULT (ARROW SHOWS DIRECTIONAL MOVEMENT)
43. REVERSE FAULT; HIGH ANGLE FAULT (ARROW SHOWS DIRECTIONAL MOVEMENT)
44. HORIZONTAL MOVEMENT IN SHEAR OR TEAR FAULT (A-MOVEMENT AWAY FROM OBSERVER; T-MOVEMENT TOWARD OBSERVER)
45. KLIPPE
46. WINDOW OR FENSTER

FIG. 4-17 Geological structure symbols

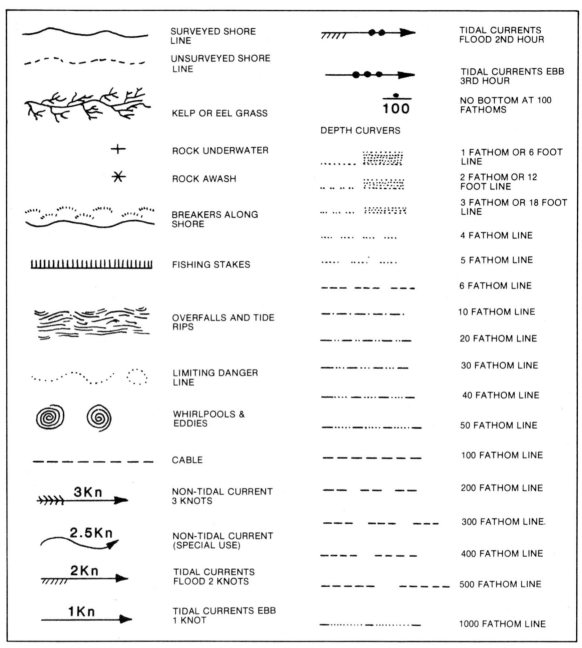

FIG. 4-18 Hydrographic and navigation symbols

given to effectively communicate the exact meaning of the drawing. These types of written information are referred to as *annotations.*

In the preparation of topos, annotations are normally placed on the drawing last. The type and amount of annotations included are generally controlled by the intent and design of the drawing. The use of annotations dictates consideration of six informational factors. These factors are lettering styles, scales and grids, measure-

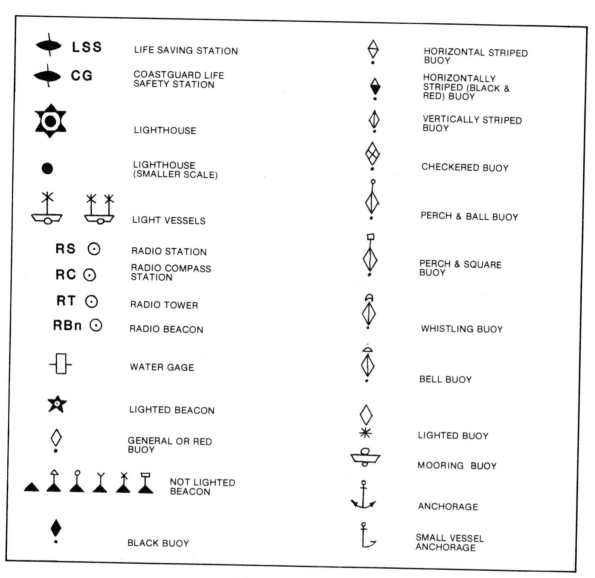

FIG. 4-18 *Continued*

ments, notations, compass points, and abbreviations.

Lettering Styles

Definite guidelines should be followed in the use of lettering styles on topographic maps and drawings. Acceptable standards dictate what type of lettering can be used to name or note the various kinds of topographic features. As shown in Table 4-3, the use of upright or vertical lettering denotes civil and land features, while inclined or italic lettering denotes hydrographic and cultural features.

The size (height) and weight (boldness) of the lettering should be in proportion to the size of the drawing and the relative importance of the information noted. It is

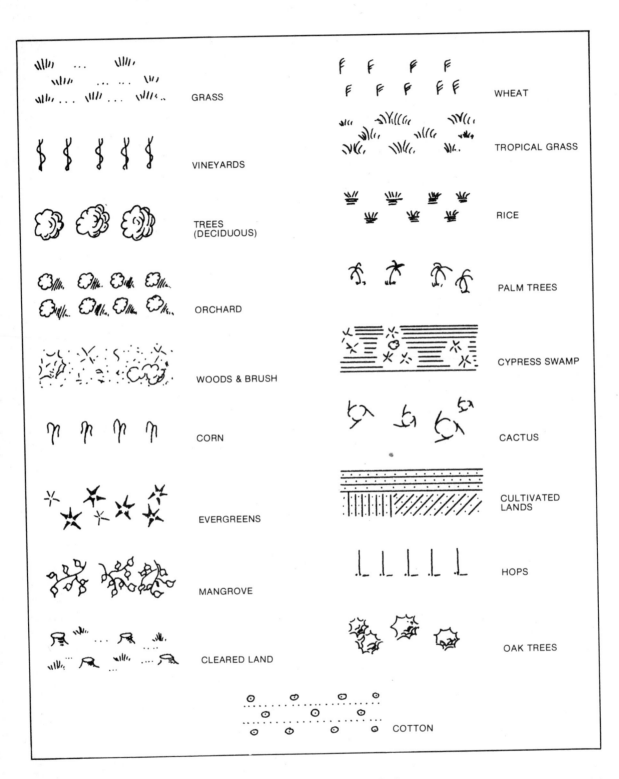

FIG. 4-19 Vegetation symbols

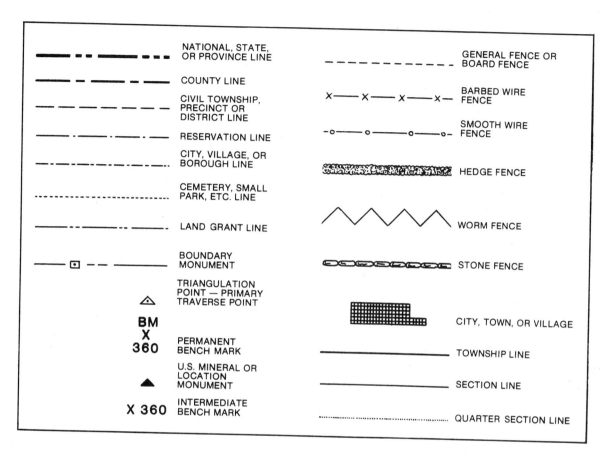

FIG. 4-20 Boundary symbols

bad practice to make the lettering so large or bold that other map symbols are covered up or obscured.

Lettering can be placed on a map drawing in one of two ways—mechanical or freehand. Mechanical lettering includes the use of mechanical lettering sets as well as transfer overlays or transfer letters (press type). This technique is widely used for the production of maps that are to be printed or published. Freehand lettering is usually limited to the production of engineering drawings such as site plans, plot plans, and other engineering topos. In some cases, however, engineering firms have found it advantageous to use mechanical lettering procedures.

Both mechanical and freehand lettering require care and consideration for placement on the drawing. It is therefore recommended that the following parameters be taken into consideration when lettering maps or drawings.

1. Names should be clearly lettered next to the designated object.

2. Lettering should not obscure other topographic symbols.

3. Names for objects by themselves should be placed close to the object, on its left or right side.

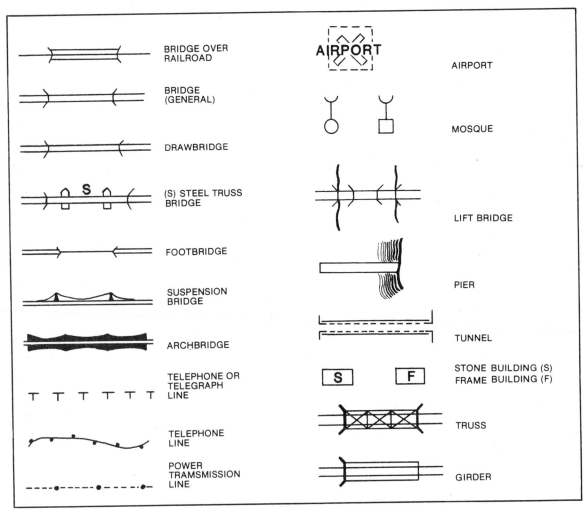

FIG. 4-21 Civil structure symbols

4. Names should be lettered parallel to the horizontal border.

5. Transportation and communication systems should be lettered close and parallel to these symbols, and as close to horizontal as possible.

6. Rivers and waterways whose widths are 2 times or more the height of the lettering should be lettered along its axis (center line).

7. On small-scale maps, when an area to be named is too small to allow any lettering, it should be referenced in a legend on the map.

8. Names of political areas (states, counties, or cities) should be lettered in the center of the area.

9. The names of elongated areas should be lettered in the direction of the longest dimension.

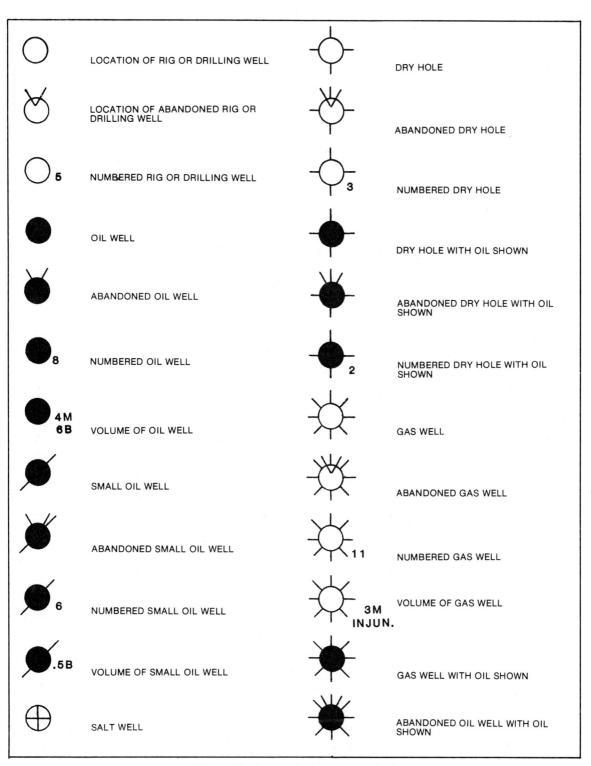

FIG. 4-22 Gas and oil works symbols

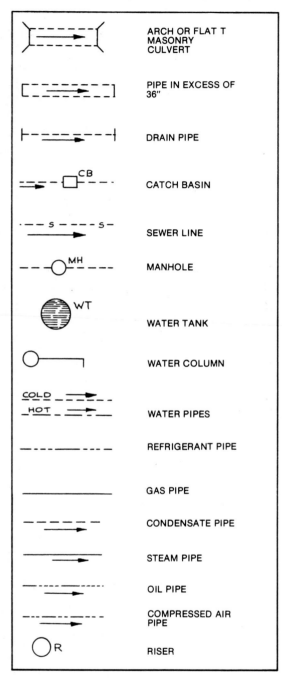

FIG. 4-23 Service system symbols

Scales and Grids

The use of scales and grids is necessary for the drawing of accurate and usable topos.

Scales are used to inform as to the proportion or ratio used on the drawing between the actual area of land map and the drawing area itself. Table 4-4 shows some common map scales. A *grid* or *graticule system* pertains to vertical and horizontal reference lines (coordinates) that are used to locate surface phenomena. For the map designer and drawer, scales and grids must be considered in union.

Scales can be shown on maps and drawings in various ways. They can be given by a statement or graphic representation, and they may be shown by the spacing of the grid system. In more subtle ways, scales are also indicated by the size of the detailing and graphic representations.

A drawing scale is an expression of distance; that is, a distance on the drawing to distance on the earth ratio. Drawing scales and distances, therefore, must always be expressed as a unit of measurement. Four of the most common ways scales may be presented on maps and topos are discussed here.

Representative fraction (RF) is a simple ratio or fraction. An example of this type of drawing scale is a scale notation of *1:63,360* or *1/63,360*. Of these two types of notations, the ratio designation (1:63,360) is preferred. The meaning of this example is that 1 unit of measure (inch, foot, centimeter) on the drawing is representative of 63,360 units (in the same unit of measure) on the surface of the earth. These scales are frequently referred to as *RF*. It must be emphasized that the units of measure on both sides of the ratio, or in the numerator and denominator, must be the same.

Verbal expressions are written statements of map distances to earth distance relationships. In our example above, RF 1:63,360 may also be expressed as 1 inch = 1 mile. Using verbal expression,

Table 4-3
Applications of Lettering Styles

Type of Information	Case	Usage	Lettering Style
Civil and Political Divisions	Upper	All states, counties, townships, large cities, and capitals	VERTICAL ROMAN
	Lower	Small towns, villages, and post offices	Vertical Roman
Hydrography	Upper	Oceans, bays, gulfs, sounds, and large lakes and rivers	*INCLINED ROMAN*
	Lower	Small rivers, creeks, ponds, marshes, brooks, and streams	*Inclined Roman*
Hypsography	Upper	Prominent topographic features	VERTICAL GOTHIC
	Lower	Small valleys, islands, hills, and ridges	Vertical Gothic
Public Works	Upper	Important railroad lines, highways, tunnels, and bridges	*INCLINED GOTHIC*
	Lower	Less important bridges, trails, roads, fords, and ferries	*Inclined Gothic*

such a map is identified as a 1-inch map. A map where 6 inches = 1 mile would be identified as a 6-inch map.

Graphic or bar scales are lines or bars that are drawn on maps and divided into various lengths. See Figure 4-24. The lengths or markings along this type of scale signify length distances along the earth's surface. When drawing graphic scales, it is common practice to subdivide one end into smaller units, enabling the user to obtain more precise measurements.

Area scales are similar to representative fraction scales, except they are ratios of surface *area*. That is to say, 1 square unit of area on the drawing is equal to 1 square unit of area on the earth's surface. Area scales can be presented in either of the following two forms: $1:63,360^2$, or *1 to the square of 63,360.*

Grids are used to locate reference points and physical phenomena on drawings. The grid, or graticule, is a systematic procedure and organization of data that identify the

Table 4-4
Common Map Scales and Their Equivalents

Map Scale	1 Inch (in) Equals	1 Centimeter (cm) Equals
1:2000	56 yards (yd)	20 meters (m)
1:5000	139 yd	50 m
1:10,000	0.158 mile (mi)	0.1 kilometer (km)
1:20,000	0.316 mi	0.2 km
1:24,000	0.379 mi	0.24 km
1:25,000	0.395 mi	0.25 km
1:31,680	0.500 mi	0.32 km
1:50,000	0.789 mi	0.5 km
1:62,500	0.986 mi	0.625 km
1:63,360	1.000 mi	0.634 km
1:75,000	1.18 mi	0.75 km
1:80,000	1.26 mi	0.80 km
1:100,000	1.58 mi	1.00 km
1:125,000	1.97 mi	1.25 km
1:250,000	3.95 mi	2.5 km
1:500,000	7.89 mi	5.0 km
1:1,000,000	15.78 mi	10.0 km

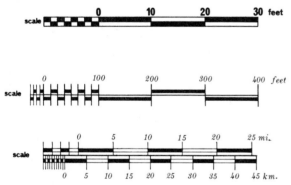

FIG. 4-24 Examples of graphic or bar scales

exact location of objects on the earth's surface, and that, in turn, can be transferred to map drawings. Hence, this system determines direction and distance.

On flat, two-dimensional planes, a rectangular coordinate system is used to identify horizontal distances (the *X value* or *abscissa*) and vertical distances (the *Y value* or *ordinate*). On a sphere, such as the earth, a much older coordinate system is used. Distance from the prime meridian is called *longitude*. Distance from the equator is called *latitude*.

Because drawings are two dimensional and spheres are three dimensional, a projection system or procedure is used to transfer map measurements. Once the spherical grid or graticule is projected to a surface, a map can be constructed. On top of the map a rectangular coordinate grid is placed, representing latitudes and longitudes.

To ensure that maps prepared with a grid system are consistent and conform to standardized procedures, two coordinate systems have been developed. The first was developed by the United States military, which has devised many grids. Two grids that have widespread use are the Universal Transverse Mercator (UTM) and the Universal Polar Stereographic (UPS). For civilian use, the United States Coast and Geodetic Survey developed the State Plane Coordinate (SPC) grid. The mapping grid system used by each state and its zones is presented in Appendix B.

Compass Points

Compass points are used to orient the map user. They indicate the direction that north is found. Figure 4-25 presents a few examples of north points. Conventional practice places north at the top of the drawing. Other orientations are used to provide a better fit of area to the map format.

Abbreviations

Topo drawers often rely on abbreviations to communicate information. Any abbreviations must be either commonly ac-

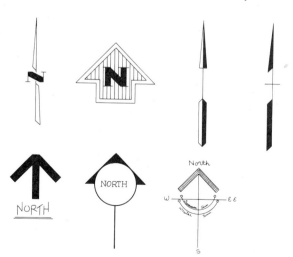

FIG. 4-25 Examples of north compass point symbols

cepted throughout the profession or defined in legends on the drawing itself. In many instances, abbreviations are used in combination with symbols. For example, gas lines are indicated by a long dash line, a *G*, and another long dash line. Table 4-5 lists some common abbreviations found on topographic drawings.

Cartographers and other map drawers have used pictorial presentations to delineate topographic features for hundreds of years. This procedure, in fact, was used in the earliest days of mapping and map drawing.

For ease of discussion, pictorial presentations are divided here into two categories: frontal (elevations) and oblique. *Frontal* presentations refer to maps that are drawn as if the individual were viewing the land formations through the frontal plane, Figure 4-26. *Oblique* presentations are those map illustrations that present topographic features as if they were viewed on an oblique (less than 90°) angle.

In both types of pictorial presentations, procedures and practices must be systematic. They must also conform to acceptable standards. These practices can be traced back to the mid-nineteenth century. The procedures used in drawing pictorial illustrations of topographic features employ the use of short, controlled lines called *hachures*. Figure 4-27 shows examples of how hachures are drawn.

These rules should be followed when drawing hachures:

1. Hachures are to be drawn in the direction of maximum slope.

73

Table 4-5
Common Abbreviations

ABAN	Abandon, Abandoned	EOP	End of Project	PI	Point of Intersection
ACC	Access	EXCA	Excavate	POC	Point on Curve
ADJ	Adjacent	EXG	Existing	POT	Point on Tangent
AGG	Aggregate			PT	Point of Tangency
ALT	Alternate	FA	Federal Aid	PP	Power Pole
APPR	Approach	FAI	Federal Aid, Interstate	PROJ	Project
APX	Approximate	FAP	Federal Aid Project		
APR	Apron			RAD	Radius
AD	Area Drain	FL	Flowline	R	Range
ASPH	Asphalt	FH	Flush Hole	RR	Railroad
AVE	Avenue	FS	Full Size	RY	Railway
AZ	Azimuth	FUT	Future	REF	Reference
				RP	Reference Point
BL	Base Line	GRN	Granite	REM	Remove
BEG	Begin	GVL	Gravel	RVS	Reverse
BOP	Beginning of Project	GN	Grid North	ROW	Right of Way
BM	Bench Mark	HW	Highwater	SCH	Schedule
BS	Back Sight	HWY	Highway	SEC	Section
BW	Barbed Wire	HOR	Horizontal	SL	Section Line
BRG	Bearing			SHO	Shore(d)(ing)
BLK	Block	Ilk	Interlock	S	South
BIT	Bituminous	INTM	Intermediate	SPL	Special
BDY	Boundary	IH	Interstate Highway	SPEC	Specification(s)
BLVD	Boulevard			SC	Spiral to Curve
BLDG	Building	INV	Invert	ST	Spiral to Tangent
BR	Bridge			SAP	State Aid Project
		JCT	Junction	STA	Station
CB	Catch Basin			SH	State Highway
CL	Center Line	L	Length	SYS	System
CM	Centimeter(s)	LMS	Limestone		
CIR	Circle	LPT	Low Point	T	Tangent
CLR	Clear(ance)	LW	Low Water	TBM	Temporary Bench Mark
CX	Connection			THK	Thick(ness)
CLL	Contract Limit Line	MAX	Maximum	TOL	Tolerance
C&G	Curb & Gutter	MH	Manhole	TR	Transom
CofA	Control of Access	MED	Median	TOPO	Topography
CO	County	MI	Mile	TN	True North
CONC	Concrete	MM	Millimeter(s)	TYP	Typical
CONST	Construct(ion)	MMB	Membrane		
COR	Corner	MIN	Minimum		

74

Table 4-5 *Continued*

CR	Creek			UC	Undercut	
CS	Curve to Spiral	NGS	National Geo-	USC	United States	
CULT	Cultivated		detic Survey	&GS	Coast & Geo-	
CULV	Culvert	NOM	Nominal		detic Survey	
		N	North			
D	Degree of Curve	NTS	Not To Scale	VERT	Vertical	
DEM	Demolish,			VC	Vertical Curve	
	Demolition	OBS	Obscure	VG	Vertical Grain	
DEP	Depressed	OPP	Opposite			
DIAM	Diameter	OA	Overall	W	West, Width	
DIAG	Diagonal			WT	Water Tank	
DIM	Dimension	PAR	Parallel	WO	Without	
D	Drain	PK	Parking	WPT	Working Point	
DI	Drop Inlet	PV	Pave(d)(ing)			
		PVMT	Pavement			
E	East	PC	Point of Curvature			
ELEC	Electric(al)	PCC	Point of Com-			
EL	Elevation		pound Curve			

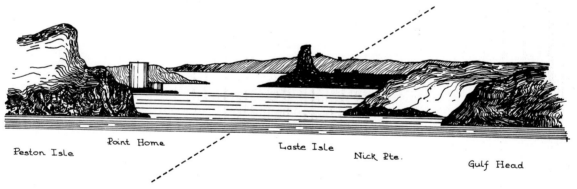

Peston Isle Point Home Laste Isle Nick Pte. Gulf Head

FIG. 4-26 Frontal pictorial presentation

2. Hachures should be arranged in rows, and not drawn as long lines down the entire slope.

3. The length of each hachure should indicate an equal vertical drop in height.

4. Hachure thicknesses should be constant, except when a form of "illuminated relief" must be presented. Hachure line thickness and hachure interspace are held constant for a given slope zone, but are different for each slope class shown on the drawing.

On many large-scale drawings, it is difficult to communicate the characteristics of a surface phenomenon with traditional hachures. Two common variants are used. One is the *illuminated hachure* — hachures are thinned or deleted on illuminated slope, assuming that the sun will shine on

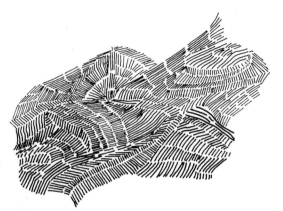

FIG. 4-27 Examples of hachures

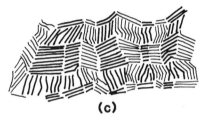

FIG. 4-28 Hachures used in various situations (**A**) Horizontal hachures (**B**) Vertical hachures (**C**) Combined

the slope at an angle. The second variant is the use of rock or cliff symbols, which can be added to drawings to enhance features such as cliffs. In some cases, modified hachures are used in combination with traditional hachures. As can be seen in Figure 4-28, this technique can be used to present such phenomena as cliffs, rock outcrops, and embankments.

4.4 OVERLAY DRAWINGS

The need for rapid data collection and processing has required, in turn, the development of new techniques and procedures to prepare and produce topographic drawings and maps. Engineers, geographers, and other professionals not only need accurate drawings, but frequently require topos with a variety of information. Such requirements translate into complex drawings, additional drawing time, and increases in drawing and map costs.

In recent years there has been a cooperative effort within the topographic field to improve the quality, accuracy, and rate of producing topos. Color is an example of how modern technology has helped in making these drawings easier to interpret. Graphic arts procedures now make it cost effective and easy to reproduce high-quality topos in multiple colors. To meet the increased demand for multicolor topos and other complex drawings, a new, sophisticated procedure has been developed and refined. This procedure is called *overlay drawing.*

Simply, overlay drawing is a process in which one base drawing or map is prepared. Next, a series of overlays (separate sheets of paper or film) are placed on the base drawing map. Drawings are then made on the overlays. The overlay drawings present information that is not on the base drawing. Applicable to either single-color or multicolor topos, the entire set of base and overlay drawings are frequently referred to as a *manuscript* or *map manuscript*.

Overlay Drawing Procedures

The actual procedures used in preparing a manuscript are quite simple and do not require the purchase of any additional drawing or drafting equipment. The steps used in overlay drawing are as follows:

1. The base drawing or map is prepared. This may be either a line drawing or a photogrammetric photomap. In either case, the base drawing should be prepared on a polyester or clear acetate film so that it will withstand frequent use under overlay preparation.

2. An overlay sheet of polyester film, clear acetate, or highly transparent paper is placed on the base drawing or map. The overlay sheet is "registered" by a reference marking system for alignment purposes. Register marks can be as simple as placing three cross markings outside the boundary of the drawing.

3. Once in place, the overlay can be drawn or scribed upon. The details drawn or lettered on should not be contained on the base drawing.

4. Additional overlays may be prepared according to Steps 2 and 3. This procedure is repeated until all overlays are completed.

Single-color Drawings

Overlay drawings are particularly advantageous when several different drawings have to be prepared for the same geographic area. For example, it may be nec-essary to provide a separate drawing for property lines, drainage systems, buildings, contours, and vegetation. Without overlay drawings, each drawing would require the development of a complete drawing from scratch. By using overlay drawing, only the specific features required per drawing have to be prepared. Figure 4-29 is an illustration of single-color overlay drawing.

Multicolor Drawings

The use of overlay drawings to produce multicolor maps is almost a required procedure. The steps and procedures are exactly the same as those discussed previously, with one minor variance. This variance is in the concept of a base drawing or map. Depending upon the type of map prepared, the base drawing may or may not be used in the reproduction process. If color is used to identify nontopographic characteristics (such as population makeup, socioeconomic characteristics, or types of soils in an area), the base map will most probably be used, since color will identify these particular features.

If color is used to identify topographic characteristics (such as contours, water systems, roads, vegetation, and political areas), then the base map will not be used in the reproduction process. It will serve as a reference map. In this situation, overlays will be placed over the base map. Each overlay will represent a topographic feature with a corresponding color. With each color represented on an overlay, the base map has no further use, since the combined colors (including black) will make up the finished product.

Figure 4-30 is an example of how overlay drawing is used in multicolor situations.

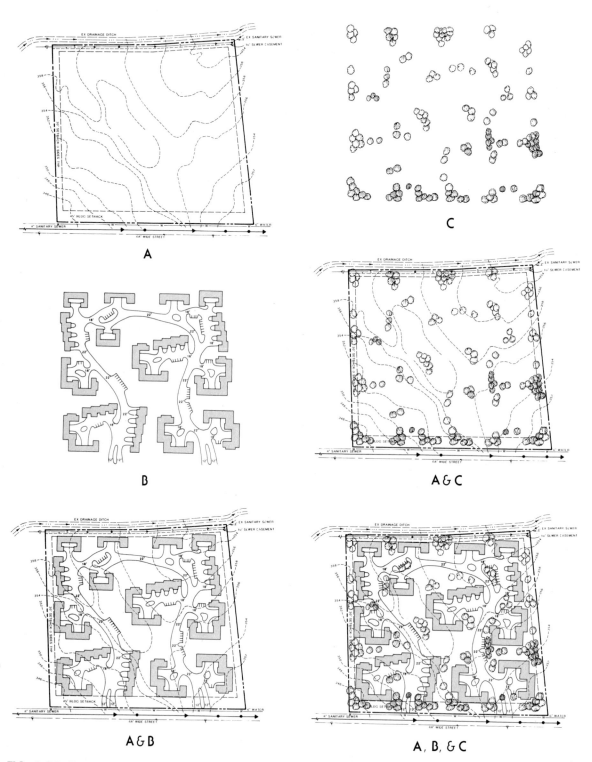

FIG. 4-29 Overlay drawings for single-color topo. Drawing **A** is the basic topo for the land, **B** is the unit and transportation layout, an **C** is the true layout.

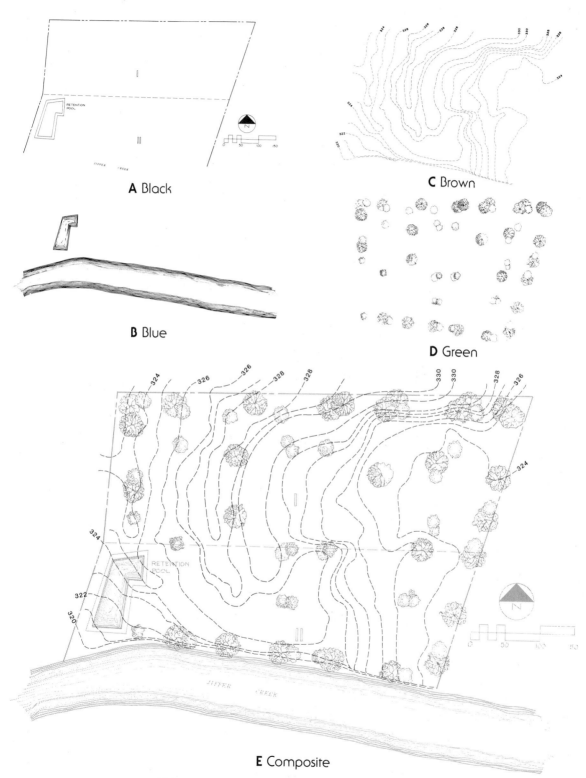

A Black

C Brown

B Blue

D Green

E Composite

FIG. 4-30 Overlay drawings for multicolor topo

4.5 SUMMARY

In mapping and topographic drawing, standard and acceptable procedures and techniques are used. The symbol is perhaps the most important convention. Point symbols are used to identify location, and line symbols are used to identify elongated and geometrically consistent phenomena. Area symbols are used to identify a two-dimensional area of land, and volume symbols are used to identify spatial variations in quantity or intensity.

Drawing topo symbols requires the consideration of several factors: composition, proportion, shading, and color. Although there are numerous symbols, they can be divided into several major categories. These are road transportation systems, railroad and waterway transportation systems, non-natural objects, map references, boundaries, surface forms, hydrographic forms, and other surface characteristics.

Detailed topographic symbols pertain to those used in small-scale maps. These symbols are more frequently found in the engineering field, where small-scale drawings are common. Unlike in other symbols, color coding is infrequently used. The major categories of symbols in this area are land and relief forms, geological structure, hydrographic and navigation, vegetation, boundaries, civil structures, gas and oil works, and service systems.

Annotations on topos are used for clarification purposes. Annotations consist of written notes, names, titles, scales, measurements, and similar data. The use of annotations, however, requires the consideration of six types of information factors. These are lettering styles, scales and grids, measurements, notations, compass points, and abbreviations.

To help clarify various types of topographic information, map drawers often use pictorial presentations. In all, there are two major classifications of pictorial presentations. Frontal presentations, or elevations, refer to map symbols that are drawn as if the individual were viewing the land formations through the frontal plane. Oblique presentations are those that present topographic features as if viewed on an oblique angle.

A technique that has been developed to save time and increase the accuracy of topo drawings is overlay drawing. This is a process whereby a base drawing or map is prepared and a series of overlays are placed on it and then drawn upon. Overlay drawing can be used for either single-color or milticolor.maps.

KEY TERMS

Abbreviation
Annotation
Area Scale
Area Symbol
Bar Scale
Base Drawing
Base Map
Boundary
Civil Structure
Color
Composition
Convention
Detailed Topographic Symbol
Elevations
Frontal Presentation
Gas and Oil Works
Geological Structure
Graphic Scale
Graticule
Grid

Hachure
Horizontal Distance
Hydrographic Forms
Illuminated Hachure
Land and Relief Form
Lettering
Lettering Style
Line Symbol
Manuscript
Map Manuscript
Map Reference Symbol
Navigation Symbol
Non-natural Object
Oblique Presentation
Ordinate
Overlay Drawing
Point Symbol
Proportion
Railroad System Symbol
Representative Fraction

RF
Road Transportation Symbol
Scale
Shading
SPC
Standards
State Plane Coordinate
Surface Form
Symbol
Universal Polar Stereographic
Universal Transverse Mercator
UPS
UTM
Vegetation Symbol
Verbal Expression
Vertical Distance
Volume Symbol
Waterway Transportation
  Symbol

REVIEW

1. What are drawing conventions, and why are they important to topographic drawings?

2. Briefly explain the differences among point, line, area, and volume symbols. Give examples of each.

3. When using and drawing symbols on topos, explain why composition, proportion, shading, and color must be taken into account.

4. Explain under what conditions color is frequently used, and for what reasons.

5. What is the primary difference between the symbols used on large-scale drawings and those used on small-scale drawings?

6. Identify the agency responsible for standardizing map symbols in the United States.

7. List the major categories of map symbols for large-scale maps, and explain the appropriate color coding for each.

8. What is an annotation?

9. Describe the system of lettering styles used on topographic maps and drawings.

10. Identify the rules that should be followed when lettering on topos.

11. What are map scales, and how are they used?

12. Explain the meaning of the following scale notation: RF 1 : 62,500.

13. What are the two forms of representation used in area scales?

14. Discuss the meaning and use of grids.

15. Explain the relationship among UTM, UPS, and SPC.

16. Why are pictorial presentations necessary in topographic drawing?

17. How are hachures used to describe surface characteristics?

18. What are overlay drawings?

19. Explain the relationship between overlay drawings and a map manuscript.

20. Describe how overlay drawing would be used in preparing a multicolor map.

ACTIVITIES

1. Draw a bar scale for both miles and kilometers for the following scales:

   a. 1 : 2000   d. 1 : 100,000

   b. 1 : 50,000   e. 1 : 250,000

   c. 1 : 63,360

2. Select a familiar land formation in your community, and sketch oblique and frontal presentations of the formation.

3. Prepare a set of overlay drawings for the drawing in Figure 4-31. Include the following colors: (a) black, (b) green, (c) brown.

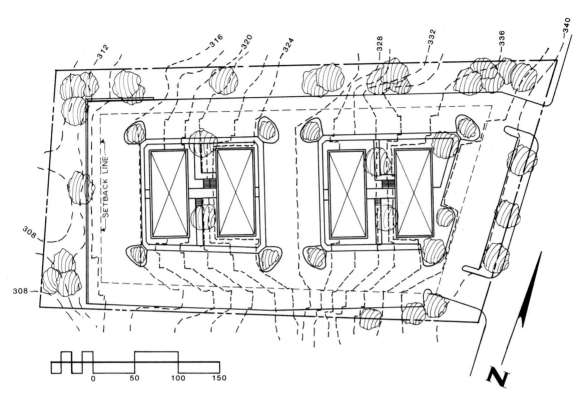

FIG. 4-31

5 CONTOUR DRAWINGS

In discussions about the drawing of topographic maps, perhaps the first image that comes to mind is the use of contours over a given land area. Because of their importance to the accurate depiction of *relief,* or land surface form, the drawing of contours is central to topographic mapping. Contours that are drawn inaccurately or with incorrect representation can make a map unrepresentative of the land in question. Such a map is, of course, useless. Thus, it is extremely important that topo drawers have a solid understanding of what contours are and how they should be used on maps and drawings.

5.1 CONTOURS

A *contour* is a line on a map drawing that shows the location of all points having equal elevation. The vertical distance between the contour lines is known as the *contour interval.* This interval is a depiction of the vertical distance between each contour line. The horizontal spacing of contour intervals, then, is determined by the slope of the land. That is, the distance will be greater as the land becomes flatter, and will decrease as the slope becomes more significant. Figure 5-1 is an example of contour lines and intervals for a map that properly translates the profile of land.

The numbers shown on the contour lines in Figure 5-1 show the elevation of each line. Most elevations are stated relative to sea level. Others, however, are stated in relation to some fixed reference point. In either case, the reference elevation is referred to as the *datum*. Hence, it is possible to have both positive (+) and negative (−) elevations; that is, elevations above and below the datum.

In a few cases, it may be necessary to show elevations relative to two datum points. This can be represented in several ways. One method involves the use of color. For another example, see Figure 5-2.

A basic question important to all contour maps is "How frequently should I note elevation levels?" The answer is dependent upon two major considerations: map scale, and land relief. The smaller the map scale, the larger the interval. The more significant the land relief, the larger the interval. Thus, one map may note elevations at every foot, while another map may note elevations at every 100 feet. In any case, the notations should be consistent throughout the drawing. These constant contour intervals are known as *index contours*.

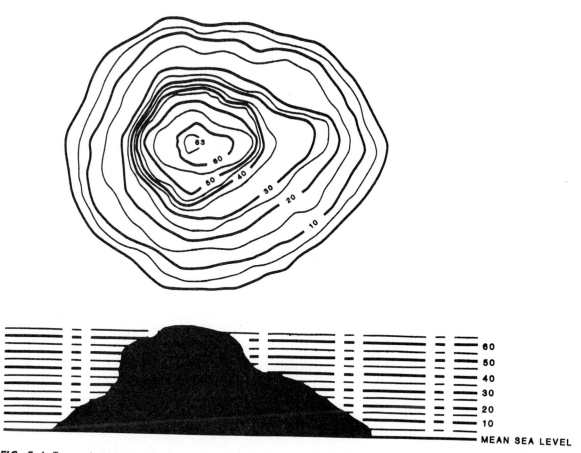

FIG. 5-1 Topo drawing with proper contour lines and intervals The profile shown below the contour is for illustration purposes only

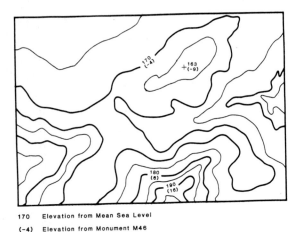

| 170 | Elevation from Mean Sea Level |
| (-4) | Elevation from Monument M46 |

FIG. 5-2 Example of technique used to show contour elevation relative to two datum points

---

5.2
CHARACTERISTICS OF CONTOURS

Contour drawings are abstract representations of real world phenomena. They are frequently misunderstood by and unintelligible to the lay public. To the trained professional, however, contour drawings are valuable tools and references.

To assure that contour drawings are prepared accurately and are representative of their respective land area, the map drawer must be cognizant of how to prepare these drawings. The map drawer must also know the characteristics of contours themselves. A number of characteristics that describe how contours should be used in map drawings are presented in this section. These characteristics are not only helpful to the map drawer, but they are also useful hints to those who read the drawing.

A contour is a continuous line which will eventually enclose itself. The line, which represents a series of points that have the same elevation, will enclose itself if the boundary of the map permits it. If the map

area is too small to permit enclosure, the contour line will end at the map boundary. See Figure 5-3.

Contours that enclose themselves should also note the elevation. An exception is if the represented area contains water. Enclosed contours with water (such as a pond or lake) will show no elevation note. See Figure 5-4.

Enclosed contours show one of two phenomena. First, the contour may represent a high peaking of land, such as a hill, mound, summit, or knoll. Second, the contour may also represent a low depression such as a closed valley, sink hole, or hollow.

The only time that contour lines will cross is where there is an overhang or an underground opening such as a cave. In such situations, the lower elevation will be shown as a dotted line or as a series of dashed lines (hidden lines).

In situations where excavation or grading work is to be done, all original contours should be drawn in as dotted or dashed lines. After excavation or grading, contours should be shown as solid lines. Appropriate elevation notations should be shown for each. In some cases the finished elevation will be lettered in a box, but this would be specified in the drawing legend. See Figure 5-5.

Contours should never be drawn across waterways or bodies of water, including streams, rivers, lakes, and ponds. Contours that are drawn to flowing water should not end abruptly at the edge of the waterway, but should turn upstream. See Figure 5-6.

Contour lines should always be drawn upgrade as they cross valleys and other low relief areas, and downgrade as they cross high relief areas.

Contour lines should be drawn evenly spaced on uniform sloping areas, and parallel and straight on plane surfaces such as roads, Figure 5-7.

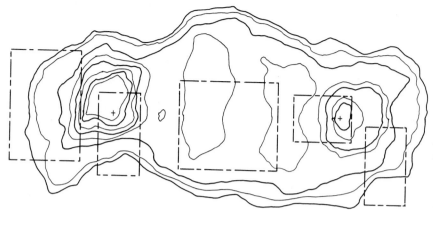

SEA LEVEL

FIG. 5-3 Conceptual illustration of how all contours eventually close on themselves

5.3
CONTOUR EXPRESSIONS

The accurate interpretation of topographic drawings is highly dependent upon the way the drawing was prepared. In addition to the skill of the drawer, there are two factors that will influence how the topo will be read. One factor is the scale of the drawing. The second factor is the contour interval used.

These two factors are especially critical where the relief is significant. Such relief is typically found in mountains, highly dissected terrain (high erosion), and glacial formations.

Improper scaling can adversely affect the interpretation of a drawing. An example is the scale chosen for a land form area comprised of steep slopes sharply incised by

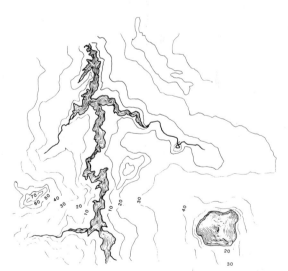

FIG. 5-4 Appropriate use of contours around bodies of water

erosion channels. If a small-scale map is drawn for this area, the contours drawn will appear to be slight bumps in roughly drawn lines. Obviously, this will not convey

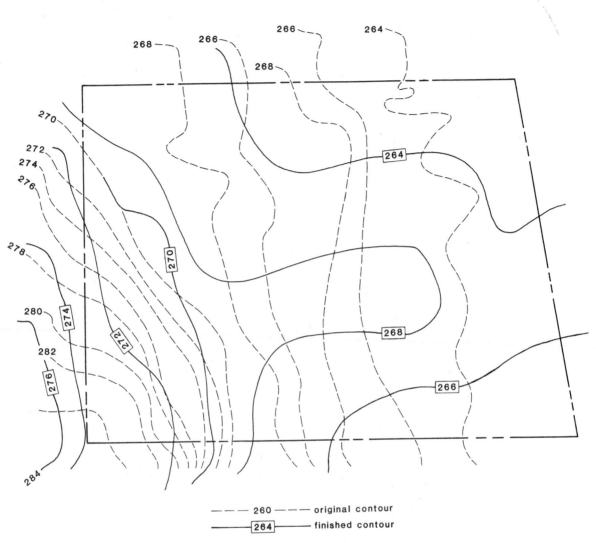

FIG. 5-5 Example of original and finished contour representation and notation

the true nature of the land. Likewise, if the contour intervals are too small (every foot or meter), the contour lines will be drawn so close together that they will merge and become unreadable. Figure 5-8 shows contour intervals spaced too close together. Figure 5-9 shows proper contour intervals.

To prepare accurate and interpretable drawings, one must pay special attention to land areas with extremes in relief. Therefore, in this section we discuss contour ex-pressions for mountains, highly dissected areas, glaciers, and flat areas.

Mountains

A *mountain* is a land formation that has steep slopes with a small summit. This is high local relief. Usually, the overall slope of a mountain is made up of other slopes that form gullies and valleys. The slopes of

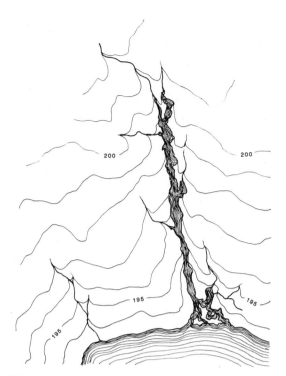

FIG. 5-6 Relationship of contours with waterways and body of water

FIG. 5-8 Contour intervals spaced too closely for scaled drawing

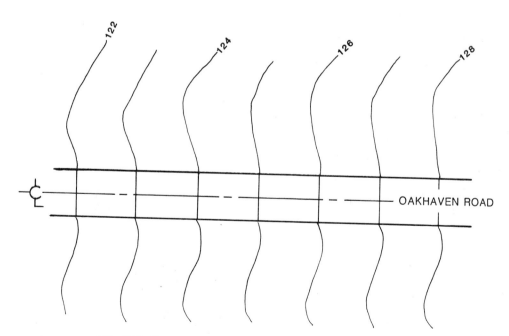

FIG. 5-7 Contour lines passing across a flat plane surface

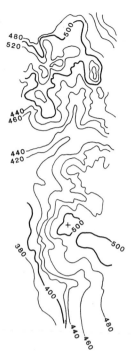

FIG. 5-9 Contours drawn at proper intervals for scaled drawing

most mountains have an average range of 20 degrees (20°) to 25 degrees (25°) from horizontal. Near the peaks, however, slopes often exceed 40°.

Mountains and mountainous areas often have sheer cliffs that rise almost 90° from horizontal. In some cases, there may even be overhangs. Geographers have generally defined mountains as formations with a minimum elevation of between 1000 feet to 3000 feet. Reliefs of less than 1000 feet are known as *hills*.

Contours used to represent mountains should be carefully drawn. Care is necessary because there is more than one type of mountain, and an incorrect representation would mislead the reader. Hence, the contour expressions used for mountains will be based upon the type of formation being drawn. In all, there are four major categories, or types, of mountains. These are folded, fault-block, volcanic, and residual mountains.

Folded mountains are easily recognizable by a pattern of parallel ranges that are separated by valleys (Figure 5-10). These formations are created as a result of the "folding" or arching of the earth's crust. They are formed in reaction to pressures caused by lateral compression.

The summits and valleys found in folded mountains are caused by erosion. Material layers that are more resistant remain as mountain ranges. Material layers that are less resistant erode away to form valleys. Most of the major mountain ranges on earth consist of folded mountains, including the Alps, Himalayas, Rockies, and Andes.

Fault-block mountains are also easily identifiable. They usually stand isolated, and are separated from other fault-block mountains by long distances (Figure 5-11). These formations frequently have one side that slopes gradually up toward the summit, while the other side is steep. The steep side is commonly called the *fault scarp*.

The fault-block mountain is formed by the vertical movement of "blocks" of the

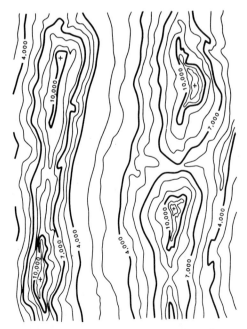

FIG. 5-10 Folded mountain topography

earth's crust, and is aligned next to fault slippage zones. These zones make possible the vertical block movement. Some fault-block mountain formations, like the Sierra Nevadas, are formed along a single fault line, while others are formed along parallel faults. The second type of formation produces long rift valleys and block-like horsts.

Volcanic mountains fall into two subcategories: intrusive and extrusive. *Intrusive* volcanic mountains are those that are formed when molten material rises near the surface and solidifies. The resulting solidified dome is then eroded by wind and/or water that exposes hard igneous rock. This phenomenon is easily recognized because the ridges and valleys that are formed have a radial pattern. See Figure 5-12.

Extrusive volcanic mountains, on the other hand, are true mountains. They are formed by the pushing up and accumulation of lava, rock, ash, and other debris through a central crater. These mountains are conical in shape (Figure 5-13) and are typically associated with the term *volcano*.

Residual mountains, unlike the other types of mountains, are formed as a result

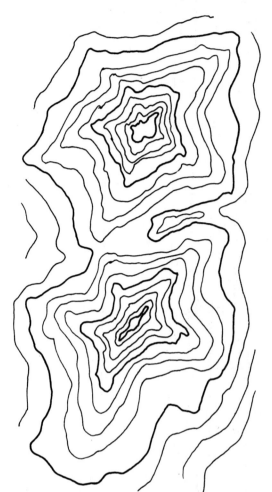

FIG. 5-12 Intrusive volcanic mountains

of long-term erosion, by wind and/or water. Over a period of time, the erosion process removes softer material or areas not brought in contact with eroding material, leaving mountainous formations. These residual mountain formations have relatively flat tops with steep sides (Figure 5-14). The smaller tops are known as *buttes*, and the larger ones are called *mesas*.

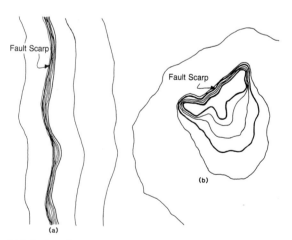

(A) Ridge line   **(B)** Single mountain formation

FIG. 5-11 Two examples of how fault-block mountains may appear

Highly Dissected Areas (High Erosion)

Land areas exposed to the forces of nature often experience high erosion. This

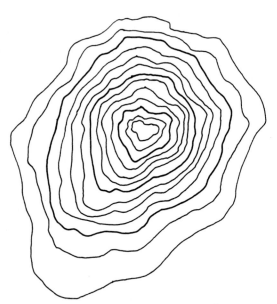

FIG. 5-13 Extrusive volcanic mountains

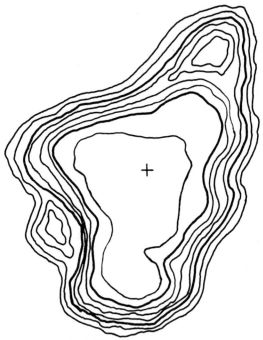

FIG. 5-14 Residual mountains

can present peculiar problems to topo drawers. Examples of two common problems are the rate of change in land relief, and the graphic representation of high-erosion areas.

High erosion can occur rapidly (over a period of days), or over long periods of time (thousands of years). Of the two situations, rapid erosion is more likely to cause problems for engineering projects. Where there is rapid erosion, existing contour drawings will prove useless, especially when large-scale drawings are used. Thus, before the project can start, a new survey must be conducted and appropriate drawings prepared.

Many curious and varied relief forms exist as a result of erosion. Therefore, no standard graphical index is used to cover all situations. However, there are procedures and concepts that must be followed when drawing contours for high-erosion areas.

To provide the drawer the background information necessary to meet and overcome the common problems, it is necessary to review two basic concepts. The first is

an overview of what erosion consists of and its cycles. The second concept is the proper representation of common erosion forms.

Erosion is the process by which land is abraded, eaten away, and shaped into gulleys, valleys, hills, and cliffs. Numerous forces cause erosion. Examples of these forces are rivers, rain, the ocean, the sun, frost, chemical action, and wind.

If no other natural force (such as continental uplifts and vulcanism) were to occur, erosion would reduce the land to a flat surface. The process of land reduction is known as the *cycle of erosion*. There are four major stages in the erosion cycle.

1. *Topographic youth* is characterized by the formation of a new uplift of land caused by crustal movements. As shown in Figure 5-15A, in this first stage the land is fairly flat, with minimal surface erosion.

2. *Topographic maturity* is where significant quantities of land have been eroded

---

92

away to produce pronounced relief forms. Notice, in Figure 5-15B, how clearly defined are the mountains, ridges, and valleys.

3. *Topographic old age* is the third stage of the erosion cycle. Where pronounced relief forms once stood, in their place is now softer and less rugged terrain. Figure 5-15C shows an example of this stage.

4. *Peneplain* is the final stage of erosion. Here, the land returns to a relatively flat appearance. The pronounced features have been worn down to rolling hills and plains. See Figure 5-15D.

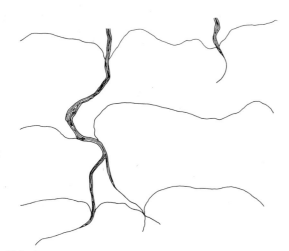

FIG. 5-15A Topographic youth

FIG. 5-15C Topographic old age

FIG. 5-15B Topographic maturity

FIG. 5-15D Peneplain

Graphical representation of erosion-producing relief should clearly describe the nature of the surface area. Drawn contours will define the stage of erosion and show the areas of water drainage. By observing a contour drawing, one should be able to learn the relative nature and extent of erosion. The critical elements in these types of drawings, then, are the characteristics and configuration of contour lines. The nature of contours must be understood by the map drawer and interpretable by the map reader. (It must be emphasized, however, that contours characterize surface forms. From them one can *deduce* the probable extent of erosion and type of agent involved.) Hence, the following contour characteristics should be known to define the nature of erosion:

- The long, continuous contours that are shown in drawings define the location and extent of erosion.

- The closer the contours appear, the steeper the terrain and more pronounced the erosion.
- Closed or enclosing contours show the location of hard, slow-eroding material.
- Closed and widely spaced contours, when used together, characterize gently rolling terrain.

Examples of these characteristics are presented in Figure 5-16. Based upon the features just described, one can clearly identify the areas of heavy erosion, and those areas of slow or minimal erosion.

Wind erosion is somewhat different from other forms of erosion, but can be confused with them. Therefore, a brief discussion is in order to describe the characteristics of wind-erosion areas.

Relief that is formed as a result of wind erosion should be easily identifiable. However, many individuals still inaccurately

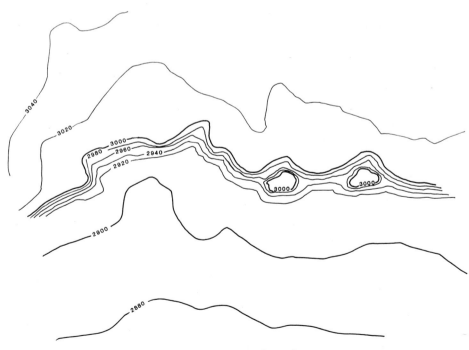

FIG. 5-16 Presentation of erosion areas

draw and interpret wind-formed contours. Figure 5-17 is a contour drawing of an area of land whose relief was formed by wind erosion. The following characteristics are seen in this example, and are common in other types of wind-erosion areas. (1) Though the area is relatively flat (nonmountainous), there is no swamp area. If a swamp area did exist, it would indicate that the relief was formed by glaciation. (2) There is no indication of water drainage areas or impeded drainage. If drainage were present, it would be an indication of water erosion. (3) There is no water shown in the depressed contour areas. (4) Closed contours show hills and mounds with relatively low elevations.

Glaciers

Glaciers are natural phenomena in which great quantities of snow are com-pressed to form ice masses. Glaciers are so enormous that the warmth of the spring and summer months is insufficient to melt them away. Thus, glaciers develop only in areas where the annual snowfall is greater than the annual melting rate.

To present glaciers and their effects in contour drawings, one must know the types and characteristics of glaciers. There are three categories of glaciers: continental, valley, and piedmont glaciers.

Continental glaciers are the largest of the three categories. Also known as ice sheets, they are of such great magnitude that they cover entire land masses. Only two con-tinental glaciers can be found today, on Greenland and the Antarctica. At one time, however, continental glaciers covered large areas of North America, Europe, Asia, and South America.

The features formed by continental gla-ciers are unique, especially on lower flat-

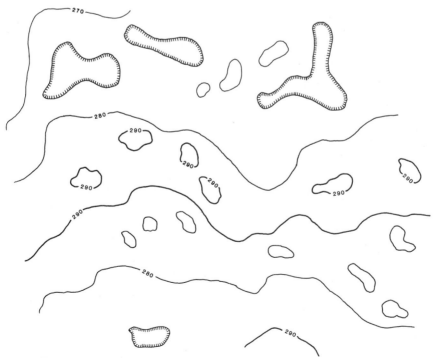

FIG. 5-17 Wind erosion for terrain containing significant sand

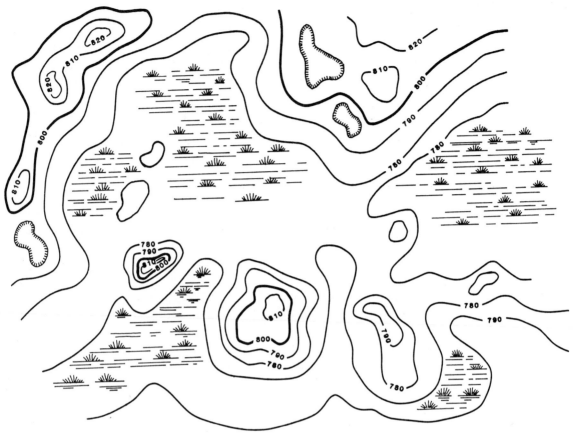

FIG. 5-18 Effects of continental glaciation

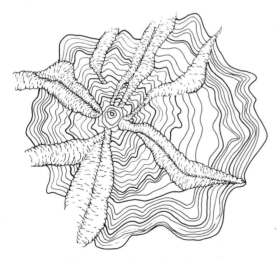

FIG. 5-19 Glaciation shown in the valleys of a mountain

lands. Figure 5-18 is an illustration of the effects of such glaciation. The characteristics of the contours of this drawing are as follows. The contours are long, smooth, and curving; this illustrates that the general topography of the land is similar to its appearance prior to glaciation. The numerous small contours are illustrations of accumulated soil, stone, and other debris carried and deposited by the glacier. (These accumulations are known as *moraines.*) Finally, the contours indicate the existence of swamps or wetlands. These features were created as a result of the natural damming of water by the glacier, which altered and stopped natural drainage.

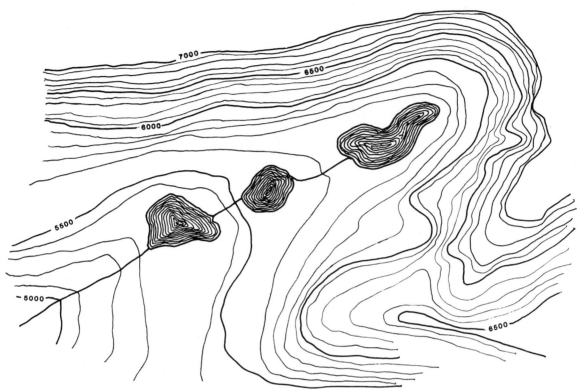

FIG. 5-20 Effects of a valley glacier

Valley glaciers are also referred to as mountain glaciers. They are so named because they are found in the high mountain tops and flow down into the contiguous valleys. These glaciers can be found in areas of high-elevation mountains that have snow on their peaks the entire year. Valley glaciers may move, like a frozen river, until they reach warmer temperatures and melt. These glaciers range in size from several hundred feet to over 70 miles in length. See Figure 5-19.

The basins formed by mountain glaciers are called cirques. They represent the valleys and lakes that remain after the glacier recedes. Figure 5-20 is a contour drawing representative of relief formed by a valley glacier. The characteristics of this drawing are as follows. The slopes of the valley are

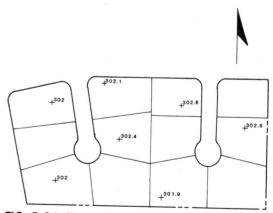

FIG. 5-21 Elevation notations for a flat area

fairly even and bottom out to a smooth U-shaped base. There is a lack of the small gullies and tributary streams that are common in other types of erosion relief. Small bodies of water (lakes) are located at the upper end of the valley. Finally, contours are pointing upstream; they are smooth and rounded. This differs significantly from a valley formed by a river, which will show sharp, pointed contours.

Piedmont glaciers are a cross between continental and valley glaciers. They consist of two or more valley glaciers that fan out, at the base of a mountain, on the low flatlands. The contours formed by piedmonts look like the continental glacier contours of flatlands, and like the valley glacier contours of valleys.

Flat Areas

Just as mountain, erosion, and glacial related expressions can pose a difficulty for the inexperienced drawer, so can flat areas. *Flat areas* are those portions of land where the elevation is so slight that for all intended purposes, the land is a plane surface. An example of this is a 4-acre development that has an elevation variance of 5 inches.

In this example, the engineer would not be interested in the drawing of contours. To be representative, the contour interval would have to be 1 or 2 inches, which would be essentially useless. Yet it may be necessary to present a topo of that site to a local building authority. For such situations that require a topo drawing, *spot,* or *point, elevations* are recommended.

As shown in Figure 5-21, spot elevations locate various points on the site and note the elevation for each. The exact location is marked with a cross (+), followed by the elevation reading. It should be realized that the elevation readings do not represent all field survey readings, just a representative sampling.

Contours drawn on topos are close estimates of the actual contour of land. Estimates are necessary because not every point on the contour can be located and measured. The contours that are drawn are plotted by points that have been positioned on the site along with its elevation readings. Since these drawings are not exact presentations of a surface area contour, they are sometimes referred to as *contour sketches.*

Due to the importance of drawing accurate contours on topographic maps, standardized procedures have been developed for contour plotting. In the topographic field, there are three recognized procedures for plotting contours, in addition to photogrammetric compilation (see Chapter 10). These procedures are using known control points, applying stadia data, and using a grid system.

Use of Control Points

The procedure used to plot contour lines by use of known control points is illustrated in Figure 5-22. Compared to the other two techniques used to plot contours, this is perhaps the least used technique. However, there are times when detailed surveys cannot be conducted, and known control points are the only available information.

Unless a sufficient number of control points is known, they can only be used for a rough estimation of the surface contour. For this reason, the control-point procedure is not used often for engineering projects. It is used, however, to map large areas of land at a small scale.

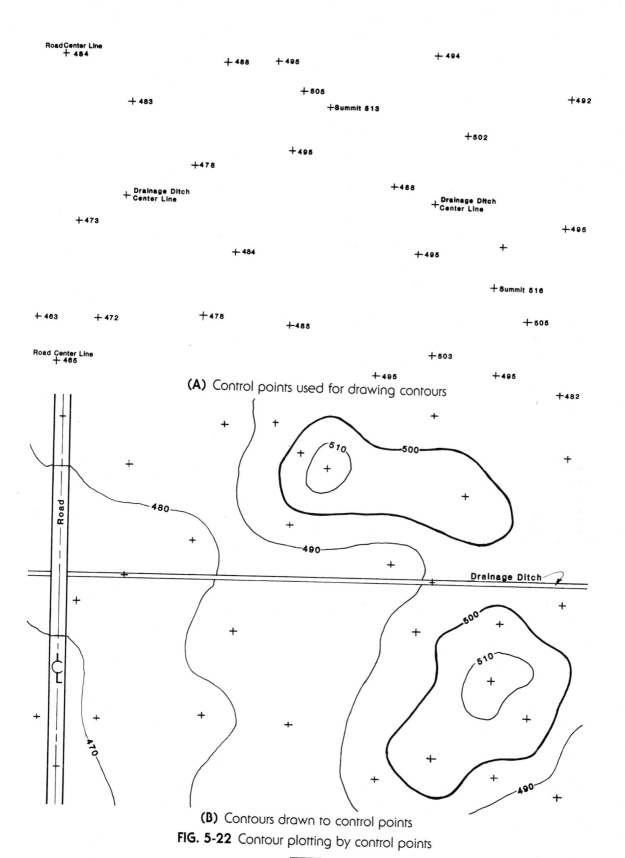

(A) Control points used for drawing contours

(B) Contours drawn to control points

FIG. 5-22 Contour plotting by control points

The process of using control points requires that the map drawer be competent in interpolating and sketching a contour system compatible with the information given. Such expertise is found in professionals who have both field surveying and drawing board experience.

Frequently, the control points available are limited to a number of key or significant land formations or constructed objects. Examples of these are summits, stream and river lines, divides, road center lines and intersections, and plane areas. Such information gives the drawer a "feel" of what the surface feature is like.

From these points, and their elevation readings, the map drawer will *interpolate* all contour lines at specific intervals. The process of contour interpolation is quite simple, but can be confusing on drawings with significant changes in relief. Consider the following example.

Points *A* and *B* are 420 feet apart, at elevations of 1373 feet and 1384 feet respectively. The problem is to interpolate the location of the 1380-foot contour.

To solve this problem, one must assume that the rate of elevation change between points *A* and *B* will be constant. By calculation (linear interpretation) we find that for every foot of rise in elevation, there will be 38.2 feet of run. Based upon this observation, the 1380-foot contour is located between points *A* and *B*, at a distance of 267.3 feet from *A* and 152.7 feet from *B*.

If this procedure is to be used to plot contours, it is recommended that the following steps be observed.

1. Interpolate and draw all closed contours. This will include all depressions and summits.
2. Interpolate the location of all contours that will cross standing and running waterways. These contours should turn upstream of running waterways.

3. Locate and draw in roads and plane areas (such as parking lots). Contours crossing these surfaces should be straight and parallel to one another.
4. Interpolate and draw all remaining contours at the appropriate interval.

Interpolation is dependent upon the careful selection of control points (marks or monuments) to locate breaks in slope, ridge lines, or stream channels. The control points should be situated along the boundary of the surface feature, and then plotted onto the map drawing. Next, the contours are interpolated between these points of known elevation, using the same procedure just described.

Applying Stadia Data

The application of stadia data for plotting contour lines is a quick and reliable method for preparing topographic drawings. The term *stadia* refers to distance and elevation measurements that have been obtained by surveying methods. Applying stadia data for contour plotting, then, refers to readings that have been recorded by surveying procedures and used as source data to draw contour lines.

By the description just given, almost every contour drawing procedure could be discussed in this section. In some cases, the use of photogrammetric data, contour tracing over point elevation maps, control points, and grid systems can all be considered examples of applying stadia data. For our discussion purposes, however, the use of stadia data is limited to a single procedure known as interval stadia data.

Interval stadia data are those measurements taken at predetermined distances (intervals) between points. These points do not necessarily follow a systematic pattern, such as used in the grid system. Because the

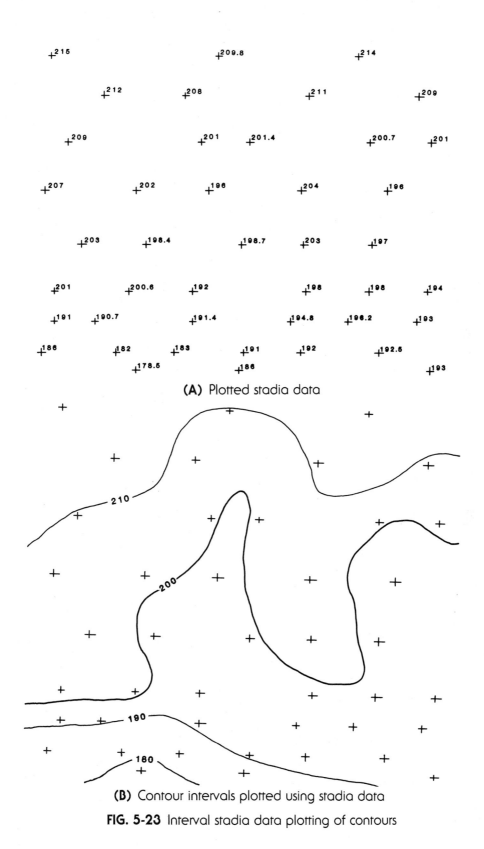

(A) Plotted stadia data

(B) Contour intervals plotted using stadia data

FIG. 5-23 Interval stadia data plotting of contours

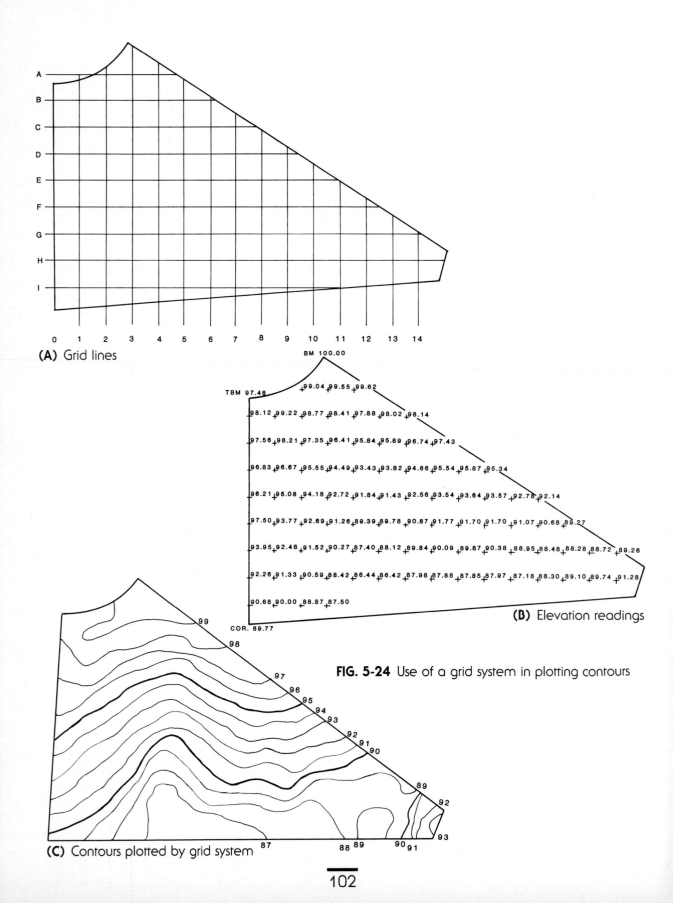

(A) Grid lines

BM 100.00

TBM 97.48

+99.04 +99.55 +99.62

+98.12 +99.22 +98.77 +98.41 +97.88 +98.02 +98.14

+97.56 +98.21 +97.35 +96.41 +95.84 +95.89 +96.74 +97.43

+96.83 +96.67 +95.55 +94.49 +93.43 +93.82 +94.66 +95.54 +95.87 +95.34

+96.21 +95.08 +94.18 +92.72 +91.84 +91.43 +92.56 +93.54 +93.64 +93.57 +92.78 +92.14

+97.50 +93.77 +92.69 +91.26 +89.39 +89.78 +90.87 +91.77 +91.70 +91.70 +91.07 +90.68 +89.27

+93.95 +92.48 +91.52 +90.27 +87.40 +88.12 +89.84 +90.09 +89.87 +90.38 +88.95 +88.48 +88.28 +88.72 +89.26

+92.26 +91.33 +90.59 +88.42 +86.44 +86.42 +87.98 +87.88 +87.85 +87.97 +87.18 +88.30 +89.10 +89.74 +91.28

+90.68 +90.00 +88.87 +87.50

COR. 89.77

(B) Elevation readings

FIG. 5-24 Use of a grid system in plotting contours

(C) Contours plotted by grid system

surveying data is recorded in a field note-book, this procedure is also known as plotting contours from stadia notes.

Figure 5-23 is an example of a contour line drawing that was plotted with interval stadia data. In preparing such a drawing, the following procedures should be observed.

Stadia intervals should be determined prior to the surveying operation. The interval selected is determined by the accuracy of the survey desired and the size of the land area to be drawn. Stadia are now recorded for all points. Control points should also be recorded. The intervals are laid out as recorded in the field notebook, and all elevations are written in. Next, by using interpolation, the contour lines are drawn in. Finally, all control points not on the same elevation as the control points are removed from the drawing.

The Grid System

The use of a *grid system* to plot contour lines is perhaps the most efficient and accurate method. As illustrated in Figure 5-24, this procedure employs a coordinate-like system of intersecting lines that are perpendicular to one another. The points where these lines intersect are the points where elevation measures are recorded.

The grid and the distance between individual lines are constant and predetermined prior to the field survey. The closer the lines, the more accurate the contour plotting.

To identify and reference each elevation point, an alpha-numeric notation system is used. This method identifies one set of lines (*i. e.*, all horizontal lines) as numbers, and the other set (*i. e.*, all vertical lines) as letters. When there are more than 26 lines along the letter side, double lettering is used. For example, if there are 28 lines along the letter side, then line 27 will be labeled AA, and line 28 will be labeled BB. In this way it is possible to refer to a single point without confusing it with another.

The steps used to plot contours by the grid system method are quite straightforward. They are as follows.

1. The perpendicular grid system is laid out to scale, using the same alpha-numeric coding used in the field.

2. The boundary of land being drawn is outlined. It should be noted that for most grid system drawings, the grid itself will extend beyond the boundaries of the tract of land, for very few contour drawings are prepared for areas of land with perpendicular boundary lines.

3. Elevation readings are taken from field notes and recorded for each point of intersection.

4. Contour lines are interpolated and drawn.

5.5 SUMMARY

Contour maps are used to show the relief of land. Contour lines show the location of all points having the same elevation, and are spaced at elevation distances known as contour intervals. To indicate the elevation of each contour line, an elevation reading is written down on contours known as index contours, at intervals of, for example, every 2, 5, or 10 feet.

Contours drawn on maps should convey specific and interpretable information. Thus, they should possess certain standardized characteristics, as follows. Contours are continuous lines that will eventually enclose themselves. Enclosed contours should always be noted with an elevation reading. Enclosed contours show either summits or depressions. Contours should only cross one another when there are overhangs or underground openings, and future changes in contours should be shown as dotted or dashed lines. Contours are not drawn across bodies of water, but are drawn upgrade and towards upstream. Finally, contour lines are drawn evenly spaced on plane surfaces.

There are a number of relief features that must be interpreted in topo drawings. Therefore, it is essential that map drawers have knowledge and skill in the drawing of contour expressions. Four major areas of contour expressions are commonly found: mountain, erosion, glacial, and flat area expressions.

The actual drawing of contour lines requires that individual contours be measured and plotted. Within the topographic field, there are three common procedures used to plot contours. These are using known control points, applying stadia data, and using a grid system. In the field of engineering, the grid system is perhaps the most common technique, while the stadia method is more common in small-scale mapping. Using known control points is reserved for inaccessible areas, and where a quick contour map is required.

KEY TERMS

Butte
Continental Glacier
Contour
Contour Interval
Contour Plotting
Control Point
Cycle of Erosion
Datum
Erosion
Extrusive
Fault-block Mountain
Fault Scarp

Flat Area
Folded Mountain
Glacier
Grid System
Ice Sheet
Interpolating
Interval Stadia Data
Intrusive
Mesa
Moraine
Mountain
Mountain Glacier

Peneplain
Piedmont Glacier
Point Elevation
Residual Mountain
Stadia
Topographic Maturity
Topographic Old Age
Topographic Youth
Valley Glacier
Volcanic Mountain
Wind Erosion

1. Explain, in your own words, the relationship between a contour and contour line.
2. What is a contour interval, and why is it used?
3. Which two factors will influence the noting of elevation levels on drawn contour lines?
4. Explain under what conditions a contour plotting would show a negative elevation reading.
5. What is the relationship between elevation readings and the datum?
6. Explain what is meant by the following statement: "A contour is a continuous line which will eventually enclose itself."
7. Describe the conditions that would show contour lines overlapping one another.
8. Explain the differences among the folded, fault-block, volcanic, and residual mountain types.
9. Give examples of how erosion can change topography over a long period of time and over a short period of time.
10. Describe the cycle of erosion.
11. List the topographic characteristics that differentiate wind erosion from water erosion.
12. Identify the major types of glaciers and their characteristics.
13. Describe how a topo would be prepared for a flat area of land.
14. Describe the difference between drawing contours by using control points and applying stadia data.
15. Develop a step-by-step procedure for drawing a contour map by use of a grid system. The procedure should start at presurveying and end in a completed drawing.

ACTIVITIES

1. Sketch a contour drawing that would show the following characteristics: negative elevation readings, a water-erosion area, an overhang, and a wind-erosion area.
2. Sketch a contour drawing showing the characteristics of fault-block mountains and spot elevations, and explain how such phenomena occur.
3. Select an area of land with known control points, and develop a contour drawing for it. If control-point data are not available, make up your own.

4. Obtain surveying data from a piece of land where a grid system was used. From this information, generate a contour drawing.

5. Using the contour drawing developed in either Activity #3 or #4, prepare a contour showing existing contours, and a proposed contour layout so that a parking area could be constructed on the land. The size and shape of the parking area is left to your discretion.

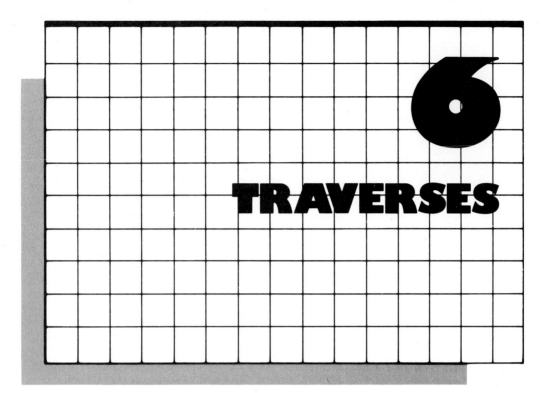

6

TRAVERSES

Traversing is a surveying procedure used to determine the direction and length of a series of lines. A *traverse* is a series of lines whose lengths are known and which are connected to one another at known angles. The traverse lines are called *courses*. The points where the courses intersect are known as *traverse stations*.

The lengths of the courses are determined by taking direct measurements of horizontal distances, slope measurements, or indirect measures based upon the surveying methods of tacheometry. The angles at the traverse stations, between the courses, are measured directly from the surveying instrument. The result of such a field survey is a series of connecting lines whose lengths and azimuths (or bearings) are

known. The lengths are called *horizontal distances*. The azimuths or bearings are either true, magnetic, assumed, or grid (these are discussed in Section 6.2).

6.1 CLASSIFICATION OF TRAVERSES

There are two broad classifications of traverses: open and closed. *Open traverses* end without closing or ending at the starting point or another known point. In comparison, *closed traverses* either close on themselves or on another known point. Of the two traversing methods, the closed is preferred, as closed traverses can easily be checked for accuracy of linear or angular measurements. To accurately check open traverses, however, is difficult, if not impossible.

Open Traverses

The unreliability of open traverses limits their use in the engineering and mapping fields. Since they cannot provide exact data, open traverses are used primarily for exploratory purposes, where estimates are sufficient. As shown in Figure 6-1, open traverses do not close upon themselves; it is thus impossible to make arithmetical checks in the field. For example, since only the starting point is known, angle summations cannot be determined, and the location of traverse stations cannot be verified.

It is important to note, however, that open traverse surveying is not as crude and nonexact as it may appear. Fairly accurate estimates can be derived by using techniques that reduce angle error. These techniques include measuring in both directions, making cellestial checks, observing magnetic bearings, and taking repeated measures. The use of these error reducing techniques will not eliminate the compounding errors, but will keep them to a minimum.

Closed Traverses

Traverses that close upon themselves are known as *loop traverses*, Figure 6-2. Be-

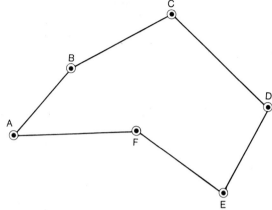

FIG. 6-2 A loop traverse

cause loop traverses form an enclosed area, it is possible to check the accuracy of the angles between the traverse courses. Field checks made on loop courses are usually limited to angular checks; hence, linear measurement checks are not made. Under these circumstances, it is easy to understand why the surveying instrument must be carefully calibrated and read to eliminate any systematic error.

Another form of closed traversing is the *connecting traverse*. Here, the traverse courses begin and end at known points, Figure 6-3. The connecting type of closed traverse is preferred to the loop type. This is because in the loop traverse only angular checks can be made, while in the connecting traverse both angular and linear measurement checks can be made. Relative to this, note that the two known points

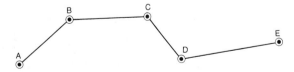

FIG. 6-1 An open traverse

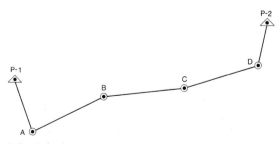

FIG. 6-3 A connecting traverse

were located by procedures as accurate as, or more accurate than, those used in the traverse survey itself.

6.2
AZIMUTHS AND BEARINGS

Horizontal angles (versus vertical angles used for elevation readings) are given according to a standard system of measure. Such measurements indicate the direction of the line or course. The direction, then, is expressed in terms of azimuth and/or bearing of a line. Azimuths and bearings are frequently shown for each boundary line about an area of land.

Azimuth of a Line

The *azimuth* of a line is the horizontal angle between it and a reference meridian. (The reference meridian is usually North, but a South meridian is sometimes used for geodetic surveys that cover large areas.) The angle is measured in a clockwise direction from the referenced meridian to the line. The direction, then, is relative to the reference meridian.

An azimuth reading can vary from 0° to 360°. An example of azimuth readings is illustrated in Figure 6-4. As shown, the North meridian is represented by the line NS (North-South) that passes through point A. The azimuth of line AB is 45°, AC is 100°, AD is 130°, and AE is 240°.

As can be surmised, a variety of reference meridians can be used. The type of meridian system selected defines the type of azimuth used. There are four types of azimuths:

- **True Azimuth**—measures from the true meridian

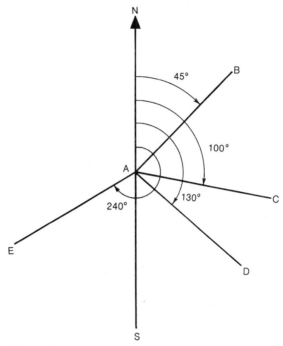

FIG. 6-4 Azimuth readings

- **Magnetic Azimuth**—measures from the magnetic meridian
- **Assumed Azimuth**—measures from an arbitrary meridian line
- **Grid Azimuth**—measures from a central meridian in a grid system

The two grid systems used in the United States are the *Lambert conformal projection* and the *transverse Mercator projection*. In each grid system, one true meridian is selected; this is known as the *central meridian*. All other North-South lines are drawn parallel to the central meridian. As shown in Appendix B, the coordinate grid systems used for each state are one of the projections described above. These systems are referred to as *State Plane Coordinate* systems.

Bearing of a Line

The *bearing* of a line is the horizontal angle relationship between it and the

North-South meridian (whichever is nearer). An added designation of East or West is also used for directional clarification. Unlike an azimuth, a bearing angle never exceeds 90°.

Bearing is used to indicate the direction of a course by specifying its angle measured from the North or South meridian, and whether that angle is measured toward the East or West. Therefore, bearings are given according to four quadrants: Northeast, Southeast, Southwest, and Northwest. For example, a course with a bearing of S 56° E identifies the direction of the line in the Southeast quadrant, 56° from the South meridian.

Figure 6-5 is an illustration of bearing readings and their relationship to azimuths. The configuration of lines in this illustration is the same as in Figure 6-4, except that the line specifications are given in terms of bearing. Line AB has a bearing of N 45° E, AC is S 80° E, AD is S 50° E, and AF is S 60° W.

Like azimuths, bearings can be categorized into four major types:

- **True Bearing** — measures from the true North-South meridian

- **Magnetic Bearing** — measures from the magnetic North-South meridian
- **Assumed Bearing** — measures from an arbitrary North-South meridian
- **Grid Bearing** — measures from a central North-South meridian

In some situations, the topographic drafter will come across the terms *record bearing* and *deed bearing*. Record bearings are measures made in reference to a previously recorded survey. Deed bearings are measures made in reference to a property deed.

Relationship Between Azimuths and Bearings

Information provided to the map drawer may be in terms of azimuths or bearings, Figure 6-6. To simplify computations and

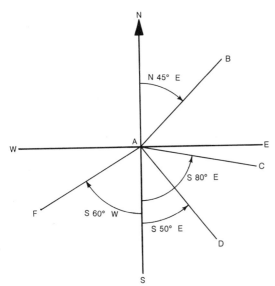

FIG. 6-5 Bearing readings

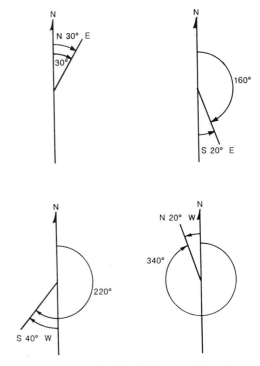

FIG. 6-6 Equivalent azimuths and bearings

the presentation of data on the drawing, it may be necessary to convert azimuths to bearings or bearings to azimuths. The conversion processes used are simple and straightforward.

The process of converting azimuth readings to bearings is easy to remember, and little practice is required to master it. The rules for converting azimuths to bearings are as follows.

1. An azimuth from North (Az_N) between 0° and 90° is in the Northeast quadrant. The bearing angle is the same as the azimuth.

2. An azimuth from North between 90° and 180° is in the Southeast quadrant. The bearing angle is calculated by subtracting the azimuth from 180° ($180° - Az_N$).

3. An azimuth from North between 180° and 270° lies in the Southwest quadrant. The bearing angle is calculated by subtracting 180° from the azimuth ($Az_N - 180°$).

4. An azimuth from North between 270° and 360° is in the Northwest quadrant. The bearing angle is calculated by subtracting the azimuth from 360° ($360° - Az_N$).

5. An azimuth from North at 0 ° or 360° has a bearing of due North; an azimuth of 90° has a bearing of due East. An azimuth of 180° has a bearing of due South, and an azimuth of 270° has a bearing of due West.

To convert bearings to azimuths, simply reverse the above process. Specifically, the rules for converting bearings to azimuths are as follows.

1. The azimuth of a line in the Northeast quadrant is the same as the bearing angle.

2. The azimuth of a line in the Southeast quadrant is 180° minus the bearing angle (180° − bearing).

3. The azimuth of a line in the Southwest quadrant is 180° plus the bearing angle (180° + bearing).

4. The azimuth of a line in the Northwest quadrant is 360° minus the bearing angle (360° − bearing).

6.3
PLOTTING TRAVERSES

Traverses can be plotted to scale accurately and reliably through a variety of plotting techniques. The procedure selected, however, will depend upon how the survey data were recorded, the field calculations made, and the preferences of the drawer. The most accurate techniques are those that employ numerical computations rather than direct field readings, since mathematical procedures are more exact.

Many traverse plotting procedures have been developed and refined by scientists and engineers. Seven traverse plotting techniques are discussed in this section. These techniques are plotting by: interior angles, distance and bearing, azimuths, deflection angles, tangents and cotangents, latitudes and departures, and rectangular coordinates.

Interior Angles

An interior angle traverse plotting is illustrated in Figure 6-7. The field data is included. In order to plot the traverse by interior angles, three sets of data must be known. These are the location of the starting point and its relationship to at least one other traverse course, the distances of the traverse courses, and the interior angles for each traverse station.

In Figure 6-7, beginning traverse station A is located relative to known point K

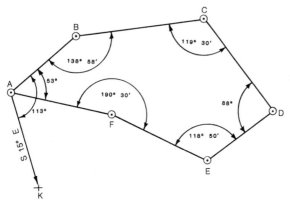

STATION	ANGLE	COURSE	DISTANCE
A	53°	A B	460
B	138° 58'	B C	705
C	119° 30'	C D	618
D	88°	D E	448
E	118° 50'	E F	583
F	190° 30'	F A	558

FIG. 6-7 Interior angle traverse plotting

(S 15° E). At point *A*, the angle *KAB* is measured by means of a supplied azimuth. Now that the direction of *AB* is known, its distance is measured off and station *B* is located. It is now possible to measure angle *FAB* and locate station *F*. The remainder of the interior angles and courses are plotted according to the data supplied.

Distance and Bearing

Plotting traverse lines by distance and bearing is perhaps the easiest method. This technique is based upon the principle of locating traverse stations relative to their bearings to other stations. In this technique, bearings are presented in two formats: bearing and back bearing. The term *back bearing* refers to the bearing of a station to a preceding station.

To plot a traverse by distance and bearing, the distance of each traverse course and the bearing/back bearing of each station must be known. In Figure 6-8, the illus-

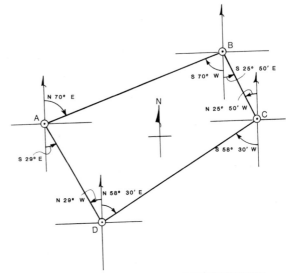

COURSE	DISTANCE	BEARING	BACK BEARING
A B	1035	N 70° E	S 29° E
B C	398	S 25° 50' E	S 70° W
C D	1003	S 58° 30' W	N 25° 50' W
D A	608	N 29° W	N 58° 30' E

FIG. 6-8 Distance and bearing traverse showing observed bearing angles

tration of a distance and bearing traverse is accompanied by the field data.

The plotting process is as follows. Beginning with station *A*, measure the bearing of *AB* and its course distance. To check the accuracy of this line, measure the back bearing to ensure that it corresponds with the field data. Repeat this procedure until the traverse is plotted.

Azimuths

Azimuth traverses present a series of lines that are related to one another by angle measurement only. Distance measures are not taken. Such traverses are used in one of two cases. One situation is when the location of a point is at such a great distance that its measurement would prove impractical. Azimuth traverses may also be used to avoid specifying azimuths for very short traverse course sections.

Two sets of data are needed to plot an azimuth traverse. One set includes the direction and distance of two known stations from which all azimuth readings are taken. The second set of data includes the azimuth readings for all traverse points from the two known stations.

Figure 6-9 shows an azimuth traverse and the recorded field data. The plotting procedure is simple to follow. For point A, the azimuth from known station L is measured and drawn. The same procedure is completed from known station M. Point A is located where the azimuth lines from L and M intersect. The same procedure is repeated for points B and C.

Deflection Angles

Deflection angles are used to indicate the direction and order of each succeeding

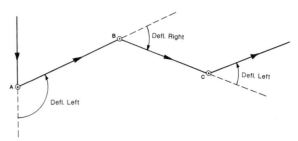

FIG. 6-10 Back and forward courses in deflection angles

traverse course. Angular measurements are made in a clockwise direction toward the forward direction. A *deflection angle* is defined as the angle between the back course and forward course as measured from a forward extension line, Figure 6-10. In other words, the deflection angle indicates the directional change of each traverse course, relative to individual traverse stations.

The use of deflection angles in surveying and plotting traverses is common in engineering projects that involve highway, railroad, and utility system construction. The

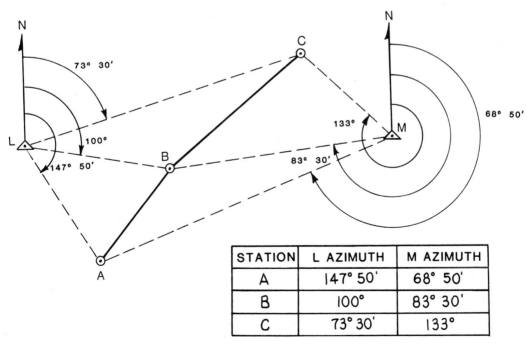

STATION	L AZIMUTH	M AZIMUTH
A	147° 50'	68° 50'
B	100°	83° 30'
C	73° 30'	133°

FIG. 6-9 Azimuth traverse plotting

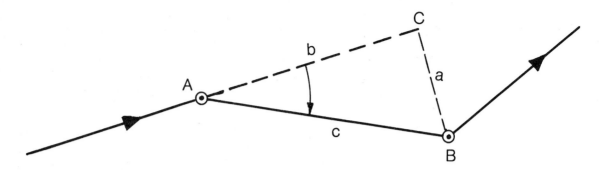

$a = 29$ $b = 61$ $c = 67$

$$\sin .5A = \sqrt{\frac{(s-b)(s-c)}{bc}}$$

$s = .5(a + b + c)$

$$\sin .5A = \sqrt{\frac{(78.5-61)(78.5-67)}{(61)(67)}}$$

$s = .5(29 + 61 + 67)$

$s = 78.5$

FIG. 6-11 Calculating half angle and deflection angle

$$\sin .5A = \sqrt{.04924}$$

$$\sin .5A = .22190$$

uses of deflection angles are threefold. First, azimuths can easily be calculated from these angles. Second, deflection angles are used to calculate circular curves in transportation systems. Third, deflection angles can be plotted easily.

$$.5A = 12° \; 49' \; 14''$$

$$A = 25° \; 38' \; 28'' R$$

When deflection angles are provided in the recorded surveying data, the location and direction of the first course must be determined and plotted. Next, each succeeding course is plotted by measuring the deflection angle (to the right or left of the station) and marking off its distance. It should be noted that deflection angles are always less than 180°. They also show the direction of turn (i.e., 32° 20′ R for a right turn, and 32° 20′ L for a left turn).

If deflection angles are not provided in the surveying data but they must be specified in the drawing, it is necessary to plot the traverse by chords. The data supplied by the survey team will include the course

or line identification, bearing, and course distance. The deflection angle and chord are calculated from this data. Instead of using the deflection angle to plot the traverse, the chord distance is the primary method of plotting.

Rather than calculating the deflection angle directly, the half angle at each station is found. A *half angle* is equivalent to the deflection angle divided by two. For an example, see Figure 6-11. In this illustration, the following formula is used:

$$\sin \frac{1}{2}A = \sqrt{\frac{(s - b)(s - c)}{bc}}$$

where: a, b, c = sides of triangle *KLM*

$$s = \frac{1}{2}(a + b + c)$$

A = angle formed by triangle *KLM* at point L; also known as the deflection angle

From the half angle, the chord distance is calculated using the following formula:

$$\text{Chord} = 2(\sin \frac{1}{2}A)(Cr)$$

where: $\sin \frac{1}{2}A$ = half angle

Cr = constant radius unit selected for a plotting; this is frequently 10 units of the scale used

The following steps are necessary for the plotting of the traverse and specification of the deflection angle. (See Figure 6-12.) First, make all necessary calculations to complete the traverse plotting table. Next, extend course *AB* for a distance equal to the constant radius. In our example, this will equal 10 feet (10 ft, or 10'). From point *B*, draw an arc with a 10' radius. Where the arc intersects the extended line *AB*, label it as point *a*.

From point *a*, draw an arc whose radius is equal to the chord distance shown for course *BC* (see the plotting table). Label the intersection of the chord radius arc and the 10' radius arc as point *b*. Now, draw a line from point *B* to point *b*. Measure the distance of course *BC* along this line and locate point *C*.

Repeat the process for each succeeding course.

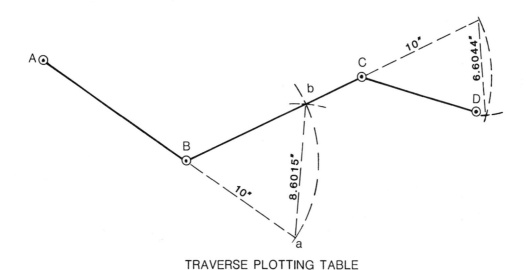

TRAVERSE PLOTTING TABLE

Course	Bearing	Distance	Defl. Angle	Sine	Cons. Rad.	Chord
A B	S50°30'E	1325				
B C		1475	59°20' L	.86015	10"	8.6015"
C D		925	41° 30' R	.66044	10"	6.6044"

FIG. 6-12 Deflection angle traverse plotting

Tangents and Cotangents

Another traverse plotting involves the use of tangents or cotangents of bearing and deflection angles. If the angle used is less than 45°, its tangent is recorded. If the angle is greater than 45°, its cotangent is recorded. With this system of plotting, a traverse plotting table should be used.

As we have mentioned, surveying data are provided to the drawer in a variety of forms. The data may include the bearing angle, the deflection angle, or both. What type of data is available will determine which plotting procedure should be selected. Therefore, depending on the type of data, one of two methods for plotting traverses by tangents and cotangents may be used. One method is the use of deflection angles. The second method is the use of bearing angles.

Deflection angle usage in the tangents and cotangents plotting method is illustrated in Figure 6-13. It is similar to the chord method of plotting traverses, but uses an *offset* instead of a chord. The procedures are as follows.

From the basic information provided, calculate the deflection angle and offset. The offset is determined by multiplying the tangent or cotangent of the deflection angle by the constant distance (10').

Next, extend line *AB* to a distance equal to the constant (10'), and label that point as *g*. From point *g*, draw a perpendicular to the right of the line, as indicated by the direction of the deflection angle. On the perpendicular, which is the offset, record the distance as shown in the plotting table and label the point as *h*.

From point *B*, draw a line through point *h*. Measure the distance of course *BC*, and label point *C*.

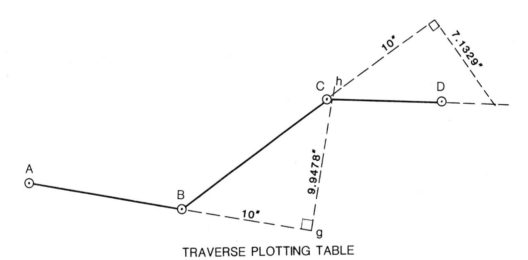

TRAVERSE PLOTTING TABLE

Course	Bearing	Distance	Defl. Angle	Tan.	Cot.	Offset
AB	S40°50'E	565				
BC		655	45°09'L		.99478	9.9478"
CD		413	35°30'R	.71329		7.1329"

FIG. 6-13 Tangents and cotangents plotting using deflection angles

Repeat the procedure for each successive course.

Bearing angle tangents and cotangents, used to plot traverses, are shown in Figure 6-14. This procedure should be used when the deflection angles are not provided and they are not needed. The steps used in this plotting technique are as follows.

Construct a square which measures, for convenience, 10 units on the map sheet, and locate the first traverse station inside the square. Its location should be tentative, since the plotted traverse might go beyond the boundaries of the square. The location of this traverse is dependent upon the shape of the traverse, the size of the paper, and the scale of the drawing.

Label each corner of the square as NE (Northeast), SE (Southeast), SW (Southwest), and NW (Northwest) respectively.

Note that the direction of the first course, AB, is N 48° E. Thus, starting at the Southwest corner of the square (start at the corner always *opposite* to the course direction), go 10 units to the East, and (10)(cot 48°) units North. Ten times the cotangent of 48° is 9.004. From the Southwest corner, heading in a northeasterly direction, draw a dotted line through the established point (10:9.004). Label this line as ab.

Line ab is now drawn in the direction N 48° E. Transfer this direction through point A and draw a line (parallel to ab), and lay off distance AB.

Repeat this procedure for each successive course.

Latitudes and Departures

The use of latitudes and departures is common in the plotting of closed traverses. The *latitude* of a course pertains to the distance that it extends in a North or South direction. Courses running in a northerly direction are said to have a plus (+) latitude. Courses running in a southerly direction have a minus (−) latitude.

In comparison, the *departure* of a course refers to the distance that it extends in an East or West direction. Courses running easterly indicate a plus (+) departure. Courses running westerly indicate a minus (−) departure.

Figure 6-15 presents course JK with a bearing of N 23° 20′ E and a distance of 805 feet, and course KL with a bearing of S 45° E and a distance of 610 feet. To calculate the latitude and departure of each course, the following formulae were used:

$$\text{Latitude} = D \cos B$$

$$\text{Departure} = D \sin B$$

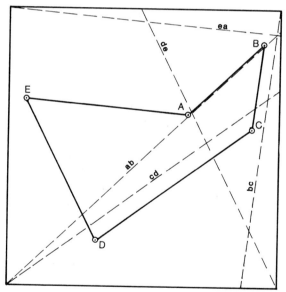

TRAVERSE PLOTTING TABLE

Course	Distance	Bearing	Tan.	Cot.
AB	48	N 48° E		.90040
BC	39	S 8° 30′ W	.14945	
CD	89	S 56° 20′ W		.66608
DE	72	N 26° 20′ W	.49495	
EA	76	S 84° E		.10510

FIG. 6-14 Tangents and cotangents plotting using bearing angles

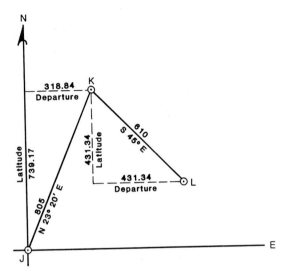

FIG. 6-15 Latitudes and departures

where: D = distance of the course

B = bearing angle

If the direction of the courses were specified in terms of azimuths from North, then the following formulae would have been used:

$$\text{Latitude} = D \cos A$$

$$\text{Departure} = D \sin A$$

where: D = distance of the course

A = azimuth angle

According to the sign quadrants illustrated in Figure 6-16, the algebraic sign (+ or −) of latitudes and departures can be determined by observing the quadrant where the bearing falls. For example, a line with a bearing of S 45° E would lie in quadrant II. The latitude would be minus (−), while the departure would be plus (+). Note that the algebraic sign corresponds to the trigonometric functions— cosine for latitudes and sine for departures.

When azimuths are used, tables with angle functions ranging from 0° to 360°, together with corresponding algebraic signs, should be consulted. If these tables are not available, it is recommended that the azimuths be converted to bearings. From bearings, algebraic signs can be assigned to latitude and departure designates.

Traverses can be conveniently plotted by latitudes and departures by the use of a *traverse plotting table*. As shown in Figure 6-17, the traverse plotting table consists

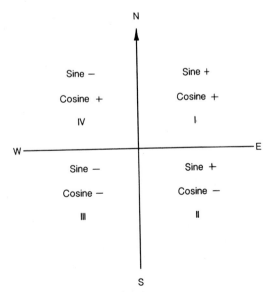

FIG. 6-16 Sign quadrants

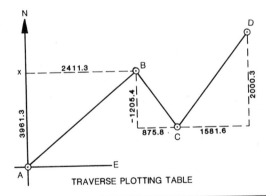

TRAVERSE PLOTTING TABLE

Course	Distance	Bearing	Sin.	Cos.	Departure	Latitude
AB	3125	N 50°30' E	.77162	.63608	+ 2411.3	+ 1990
BC	1490	S 36° E	.58779	.80902	+ 875.8	− 1205.4
CD	2550	N 38°20' E	.62024	.78442	+ 1581.6	+ 2000.4

FIG. 6-17 Plotting traverses by latitudes and departures

of the basic surveying data plus the latitudes and departures for each course. To plot the survey, the following procedure should be used.

First, locate the appropriate position for station *A*. Draw a meridian through *A*. All latitudes will be drawn parallel to this meridian, and all departures will be drawn perpendicular to it.

To locate station *B*, measure the distance of the latitude for course *AB* in the correct direction. Label that point as *x*. From point *x*, draw a line through it and perpendicular to the meridian.

Measure the distance of the departure for course *AB* in the appropriate direction, and label that point *B*. Finally, connect points *A* and *B*.

Repeat the procedure for each succeeding course. Note that it is not important whether the latitude or departure is plotted first. Just be sure that the latitude and departure lines are drawn perpendicular to one another.

Rectangular Coordinates

Traverse plotting by rectangular coordinates is one of the most accurate methods. This procedure is based upon the premise that a closed traverse is being plotted. It is accurate because any error in plotting stations or distances can be easily observed and corrected.

There are a number of advantages to using rectangular coordinates. The three most significant advantages are discussed here. First, an error in plotting does not affect the plotting of succeeding stations. Each station is plotted independently from the others, thus making it impossible to compound errors. Second, as soon as a station is plotted, its accuracy can be checked by measuring (to scale) its distance from the preceding station. A third advantage is

that the appropriate size of the mapping paper, and the drawing boundaries, can be determined easily by examining the coordinates.

The perpendicular coordinates *X* and *Y* can be drawn arbitrarily or drawn to correspond to the meridian. If the coordinate system corresponds to the meridian, then latitudes and departures can be used to plot each point. On the other hand, if an arbitrary grid is drawn, each point is plotted according to its *X* and *Y* distances. Of the two drawing coordinates, the meridian grid is used more frequently for small-scale drawings, while the arbitrary grid is commonly used in large-scale drawings.

The procedure used to plot traverses in an arbitrary coordinate system is based upon distances from each grid line. Figure 6-18 is an illustration of this procedure.

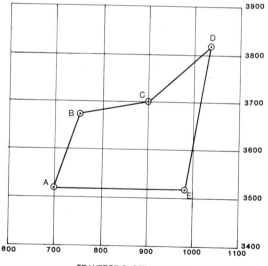

TRAVERSE PLOTTING TABLE

Station	X-Coordinate	Y-Coordinate
A	700	3519
B	750	3672
C	800	3700
D	1034	3819
E	981	3519

FIG. 6-18 Rectangular coordinate plotting

The steps that should be followed are presented here.

First, draw a series of grid lines so that each set is perpendicular to the others. The lines on each coordinate (X and Y) should be spaced at constant intervals (e. g., 10, 50, 100, or 500 feet). The actual spacing interval will depend upon the scale of the drawing.

Label each grid line with its proper distance designate on the X and Y coordinate, and plot each station according to its distance from the nearest grid line. Distances are obtained from the traverse plotting table. Check the accuracy of each plotted station by measuring the distance between it and the preceding station.

Repeat the procedure for each succeeding station.

6.4 SUMMARY

Traversing is a surveying procedure used to determine the direction and length of a series of lines known as courses. Each point where the courses meet is known as a station. The lengths of the lines are referred to as horizontal distances because they lie on a horizontal plane, as compared to vertical distances, which measure elevations.

There are two major categories of traverses. Open traverses start and end at two separate locations. Closed traverses can be either loop or connecting. Loop traverses close on themselves, while connecting traverses start and end at known locations.

Directions of traverse courses are given in terms of horizontal angles. These angles can be given in terms of azimuth and/or bearing. Azimuth readings range from 0° to 360°, while bearings are always less than 90° and give indications as to the eastern and western direction of the line. At times it may be necessary to convert azimuths to bearings, or vice versa. In such cases, a standard conversion process must be used.

Traverses can be plotted by using a variety of techniques. These include plotting by interior angles, distance and bearing, azimuths, deflection angles, tangents and cotangents, latitudes and departures, and rectangular coordinates.

KEY TERMS

Azimuth	Closed Traverse	Deflection Angle
Azimuth Angle	Connecting Traverse	Distance
Back Course	Cosine	Forward Course
Bearing	Cotangent	Grid
Bearing Angle	Course	Half Angle
Chord	Deed Bearing	Horizontal Distance

Interior Angle
Lambert Conformal Projection
Latitudes and Departures
Loop Traverse
Meridian

Offset
Open Traverse
Quadrant
Record Bearing
Rectangular Coordinate

Sine
Tangent
Transverse Mercator Projection
Traverse
Traverse Station

REVIEW

1. Explain the differences and relationships among traverses, courses, and stations.

2. Why are traverse distances said to be horizontal?

3. What is the difference between an open and a closed traverse? Of the two types of traverses, which is more advantageous to use, and why?

4. Explain the meaning of an azimuth of a line, and explain how azimuths are recorded.

5. What is the bearing of a line, and how does it differ from the azimuth?

6. Explain the difference between true bearing and azimuth, and magnetic bearing and azimuth.

7. Convert the following azimuths to bearings:

a. $Az_N = 87°$ c. $Az_N = 132° 14' 3''$ e. $Az_N = 287° 34'$
b. $Az_S = 32°$ d. $Az_S = 194° 24'$ f. $Az_S = 114° 3'$

8. Calculate the deflection angle at point A, if side $b = 32'$ and side $c = 23'$, and determine the chord distance.

9. Explain the difference between a latitude and departure, and discuss when each would have a plus or minus value.

10. Why is the use of rectangular coordinates one of the most accurate methods for plotting traverses?

1. Plot a closed traverse based upon the following data provided by a surveying team. The clockwise interior angles start at station *A*.

Station	From	To	Interior Angle	Course	Distance
A	E	B	64° 24′	AB	616
B	A	C	206° 35′	BC	691
C	B	D	64° 54′	CD	783
D	C	E	96° 39′	DE	971
E	D	A	107° 34′	EA	678

Note: The bearing of course AB is N 32° E.

2. Using the distance and bearing method of plotting, plot the following traverse course.

Course	Distance	Bearing
AB	453	N 26° 17′ E
BC	279	N 39° 18′ W
CD	422	S 80° 21′ E
DE	483	N 47° 28′ E
EF	392	S 8° 28′ W
FA	886	S 56° 27′ W

3. Plot the following traverse by using the azimuth plotting procedure, starting with point *K* and moving in a clockwise manner.

Course	Az$_N$	Distance
KL	59° 30′	125
LM	110° 10′	240
MN	270°	178

4. Based upon the following data supplied by the surveying team, calculate the deflection angle of each course and plot it by use of the deflection angle data.

Course	Bearing	Distance	Deflection Angle
PQ	N 30° E	200	
QR	N 55° 10′ E	400	
RS	N 80° 24′ E	525	
ST	N 42° E	325	
TU	S 68° 14′ E	762	

5. Using the tangents and cotangents method for plotting traverses, complete the following table and plot the traverse.

Course	Distance	Deflection Angle	Tangent	Cotangent	Offset
AB	200	0°			
BC	150	28° 14′ R			
CD	197	30° 32′ R			
DE	390	40° 20′ L			
EF	202	33° L			

6. Using the tangents and cotangents method, complete the following table and plot the traverse.

Course	Distance	Bearing	Tangent	Cotangent
FG	802.6	N 89° E		
GH	1095.7	N 24° 21′ E		
HI	1464.1	N 65° 52′ W		
IJ	882.0	S 43° 35′ W		
JK	852.0	S 58° 19′ E		
KL	524.4	S 3° 53′ W		

7. Plot the following traverse by using the latitudes and departures method. Complete the table, and indicate the algebraic sign of each latitude and departure.

Course	Distance	Az_N	Latitudes	Departures
MN	933	130° 15′		
NO	1274	65° 36′		
OP	1702	335° 23′		
PQ	1025	264° 50′		
QR	992	162° 56′		
RS	612	225° 9′		

8. Using a rectangular coordinate system, plot the following traverse and measure the distance of each course as a check for accuracy.

Station	X	Y	Course	Distance
A	500.00	820.07	AB	447.24
B	693.01	1223.52	BC	427.17
C	1095.47	1080.33	CD	400.00
D	1502.21	1028.45	DE	583.09
E	1255.77	546.29	EF	316.22
F	942.97	500.00	FA	546.50

The use of profiles is an important, yet sometimes overlooked, element of topographic and map drawing. *Profiles*, frequently used in various types of engineering projects, are defined as drawings that show the continuous vertical cross section of the earth's surface. All vertical section drawings are called profiles, regardless of the direction of the cross section.

7.1
PROFILE LEVELING

The purpose of *profile leveling* is to determine the elevational characteristics of the earth's surface along a particular course.

Before designing and constructing highways, utility systems, railroads, sidewalks, and waterways, existing profiles must be developed. Profiles determine the amount of earth moving operations that will be necessary for construction.

The route of a profile course may take one or more of the following three forms:

- **Straight Line** — for a short path or sidewalk.
- **Broken Line** — for utility systems such as transmission, sewer, and water lines.
- **Series of Straight Lines Connected by Curves** — for transportation systems such as railroads, highways, or canals.

The data used for plotting profiles are usually obtained in the field by surveying teams. This is the process of profile leveling. The data are normally recorded as profile level, or leveling, notes, and are passed along to the drafter for plotting.

A *plotted profile* can be described as a graphical presentation of a series of vertical surfaces intersecting one another. When presented, these plottings are to a larger vertical scale than the horizontal scale. The purpose of the scaling difference is to exaggerate or emphasize the difference in elevation. The typical vertical to horizontal scale ratio is 10:1. Note that the horizontal drawing is referred to as the *plan,* while the vertical drawing is called the *profile.*

Figure 7-1 delineates the principles of profile leveling and plotting. Table 7-1 presents the corresponding field notes for the illustrated profile. The procedures used by surveying teams for profile leveling are discussed in this section.

In the first step of the procedure, stakes, or markers, are set up at regular and consistent intervals. The interval will range from 25 to 100 feet, depending upon the regularity of the ground. In our example, the interval is 50 feet.

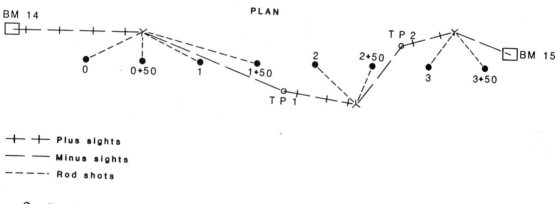

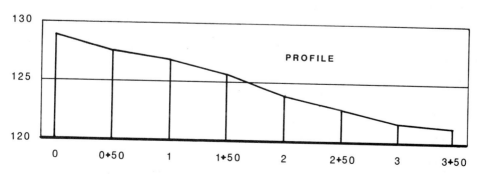

FIG. 7-1 Principles of profile leveling and plotting

Table 7-1
Profile Leveling Field Notes

Station	+	HI	−	Rod	Elevation
BM 14	2.6	133.1			130.5
0				4.2	128.9
0 + 50				5.6	127.5
1				6.2	126.9
1 + 50				7.7	125.4
TP 1	3.7	132.9	3.8		129.8
2				8.9	124.0
2 + 50				10.0	122.9
TP 2	6.0	132.6	6.4		126.6
3				10.9	121.7
3 + 50				11.2	121.4
BM 15	___		10.4		
	12.2		20.6		122.2

Check: 130.5
      + 12.3
        142.8
      − 20.6
        122.2

A 100-foot interval point is called a *full station*. Points between stations are numbered as *pluses*. For example, a point that is numbered 6 + 32.5 is 632.5 feet from station *0*.

Station *0* is the beginning point of the survey. It is always positioned relative to a known *bench mark* (BM). By using a known bench mark, the surveying team will have reference to its exact location and elevation; hence, it will be possible to arithmetically check elevation readings.

The next step is to obtain the elevation at station *0*. The level is set near station *0*, and a plus reading of 2.6 is taken on BM 14. Added to BM 14's elevation of 130.5, the *height of instrument* (HI) is calculated to be 133.1.

The rod is then read on station *0*. The 4.2 is known as a *rod reading* or *rod shot*. The rod reading is subtracted from the HI to obtain the elevation at station *0*, which is 128.9.

The notation *TP* refers to a *turning point*. A turning point is established when a rod reading exceeds 150 feet or is obstructed from observation. In our case, the first turning point is established at location TP 1.

A *minus* reading is taken on each TP and subtracted from the HI to give the elevation at that point.

Finally, the profile leveling should be terminated at another bench mark. As shown in Table 7-1, the last bench mark reading is used to check the accuracy of the elevation readings.

7.3
GRADE LINES

The term *grade* is loosely used in surveying, engineering, and topography. However, it is commonly agreed that a grade is used to describe the elevation of future or finished construction. In Figure 7-2, the irregular line represents the original profile of an area of land. The *grade line* is the line to which this profile is to be constructed by grading operations, Figure 7-3.

The rate of grade is known as the *gradient*. The gradient is the rate of change in elevation. It is specified as a ratio of elevation change to horizontal distance (see Figure 7-4). For example, if a road sloped

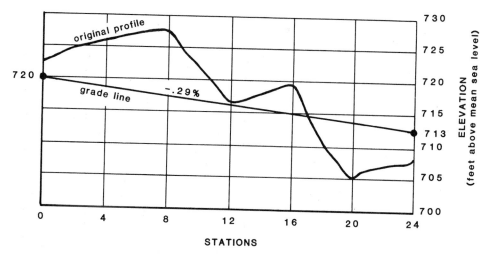

FIG. 7-2 Profile with a grade line

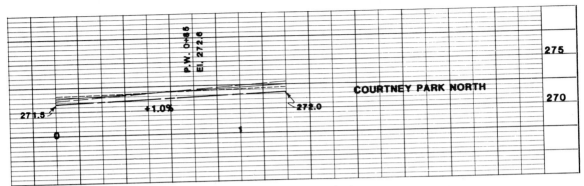

FIG. 7-3 Road grade line

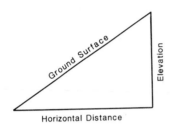

$$\text{Gradient = Rate of Change} = \frac{\text{Elevation Difference}}{\text{Horizontal Distance}}$$

FIG. 7-4 Gradient

upward 0.5 foot in a horizontal distance of 100 feet, the gradient would be +0.005, or +0.5%. If the road sloped downward, the gradient would be −0.005, or −0.5%.

The terms *fill* and *cut* are frequently used in relation to the grade. When the grade is higher than elevation, as shown in the profile, a notation to fill is written at that point. For example, F 5'-8" is a notation to fill 5'-8". When the grade is below the profile elevation, a cut notation is used, such as C 2'-10" for cut 2'-10". These terms are used only in reference to distance from grade, and not for excavation or embankment specifications.

Depending upon the type of engineering project, the grade line will pertain to different slopes. For example, consider the following situations:

1. *For highways and streets,* the grade line is the finished vertical cross section at the center line.

2. *For construction of structures,* the grade line is used to represent the subgrade.

3. *For railroads,* the grade line represents the location of the base rail.

As can be surmised, engineers use profiles mainly to establish a grade line. This is a critical process, for the cost of a project can increase or decrease in relationship to the amount of excavation or grading required.

7.4
PLOTTING PROFILES

The technique selected to plot a profile depends upon the type and form of information available. Two general methods are used, however, to plot profiles. One method involves the use of profile leveling field notes, while projection from existing contour drawings is used in the second method.

As previously mentioned, it is common to find both the plan and profile on the same drawing sheet. In these situations, the plan should appear at the top, or over the profile. This procedure is typically used on engineering projects involving highways and utility systems.

Profile Leveling Notes

The use of field profile leveling notes is the method of plotting profiles used most frequently. An aid commonly employed in this procedure is *profile paper*. Profile paper has horizontal and vertical lines that are printed in light green, blue, or orange. As can be observed in Figure 7-5, the vertical lines are more widely spaced than are the horizontal lines. This spacing style is used to emphasize the difference between vertical and horizontal measurements. The actual increment of measurement (*i. e.*, 5 or 10 feet) will be determined by three major factors: the change in elevation, the horizontal distance, and the work requirements.

If only one profile copy is required, a heavyweight paper is used. When re-productions must be run, vellum-grade profile paper is used.

Figure 7-6 shows a profile plotting from field profile level notes. The actual process of plotting is quite simple. It consists of transposing the information from the field notes to the profile paper. Briefly, the process is as follows.

Appropriate horizontal and vertical scales are established. In our example, each horizontal line represents 1 foot, as does each vertical line. Taking the information shown in the field notes (see Table 7-2), the elevation for each foot, beginning with station *0*, is plotted.

Once every vertical measurement is plotted, the points should be connected with a smooth, continuous line. Finally, all necessary notes and specifications are added.

Projection from Contour Drawings

A less common method of plotting profiles is by projecting from an existing contour drawing. This procedure is the only practical alternative available when a profile leveling survey has not been conducted. The only prerequisite is an existing contour map drawing.

Figure 7-7 is an illustration of plotting a profile from an existing contour drawing. This plotting procedure can be drawn on profile paper or plain drawing paper. For our example, profile paper was used. The plotting steps are as follows.

The contour drawing is attached or reproduced onto the profile paper. The contour drawing is then labeled the plan. An appropriate vertical scale is selected. The horizontal scale is already determined by the contour drawing.

Next, the location of the cross section cutting plane line is marked on the contour drawing. (See line *AB*.) Where cutting

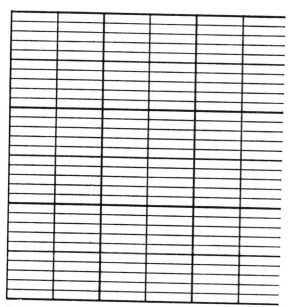

FIG. 7-5 Profile paper

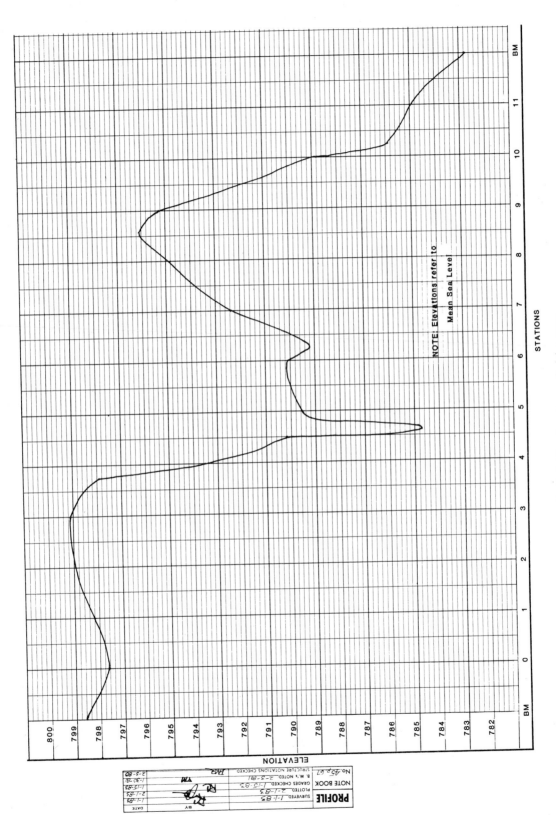

FIG. 7-6 Plotting from field notes

Table 7-2
Profile Leveling Notes

Station	+	HI	−	Elevation	Remarks
1983-1-1. Profile, Preliminary Base County Road. State. Location Notes, Field Book 45, p. 67.					
BM	4.18	802.7		798.6	Spike at base of granite rock
0			5.2	797.6	55′ left of sta. 0
1			4.6	798.2	
2			3.9	798.9	
3			3.8	799.1	
3 + 70			4.9	797.9	
4			9.2	793.6	
TP 1	1.21	794.6	9.3	793.4	Stake near sta. 4
4 + 50			4.4	790.2	
+ 63			9.9	784.7	
+ 77			5.3	789.3	
5			5.0	789.6	
5 + 65			3.9	790.4	
6			4.5	790.1	
6 + 25			5.4	789.2	
7			2.2	792.4	

Table 7-2, *Continued*

1983-1-1. Profile, Preliminary Base County Road. State. Location Notes, Field Book 45, p. 67.					
Station	+	HI	−	Elevation	Remarks
TP 2	3.67	797.2	1.08	793.6	Over white oak stump, near sta. 7
8			2.4	794.8	
8 + 65			1.1	796.1	
9			1.9	795.3	
10			8.4	788.8	
10 + 35			11.3	785.9	
11			12.2	785.0	
BM			3.4	782.8	Spike at base of fence post, 14′ left of sta. 11

plane line *AB* cuts across each contour line, that point of intersection is projected to the profile. The elevation of the intersecting contour should correspond to the same profile elevation. Intermediate points which have no contour and no point of inter-section should be interpolated by scaling proportionate distances, and projected to the profile.

Finally, the projected points are connected with a smooth continuous line, and any required specifications are noted.

7.5 SUMMARY

Profiles are used by engineering companies to show the continuous vertical cross section of the earth's surface. To obtain a profile, a surveying procedure known as profile leveling is conducted. The data are recorded in a field notebook and passed along to the drafter for graphical presentation.

During the field procedures used in profile leveling, care must be taken to record accurate elevation readings for each station. A way to check these readings is to use an established bench mark for locating the first and last station and their elevations. Other considerations in the field are station intervals, height of instrument, and turning points.

An integral part of determining a profile is the establishment of grade lines. These lines are used to describe the elevation of future or finished construction. The rate of grade is referred to as the gradient, and is specified in terms of plus and minus readings. Other considerations are fill and cut notes, which identify the relationship between the profile and grade line.

Two methods are used for plotting profiles. The method used most frequently is plotting from profile leveling notes. Another technique is projecting from an existing contour map drawing. To aid in the plotting process, profile paper is frequently used.

KEY TERMS

Bench Mark
Full Station
Grade
Grade Line
Gradient
Height of Instrument

Minus
Plan
Plus
Profile
Profile Leveling
Profile Paper

Rod Reading
Rod Shot
Station
Turning Point

REVIEW

1. What is the principal use of profiles for engineers, and what is the importance of profiles to the financial success of a construction project?

2. Describe profile leveling and its relationship to plotting profiles.

3. What is the relationship between the plan and profile in a drawing?

4. Briefly explain the significance of the following during the field surveying process:

    a. Bench mark   c. Full station

    b. Turning point   d. Rod reading

5. Explain the difference between a plus and a negative gradient.

6. What is profile paper, and what is its use in plotting profiles by profile leveling notes and contour projection?

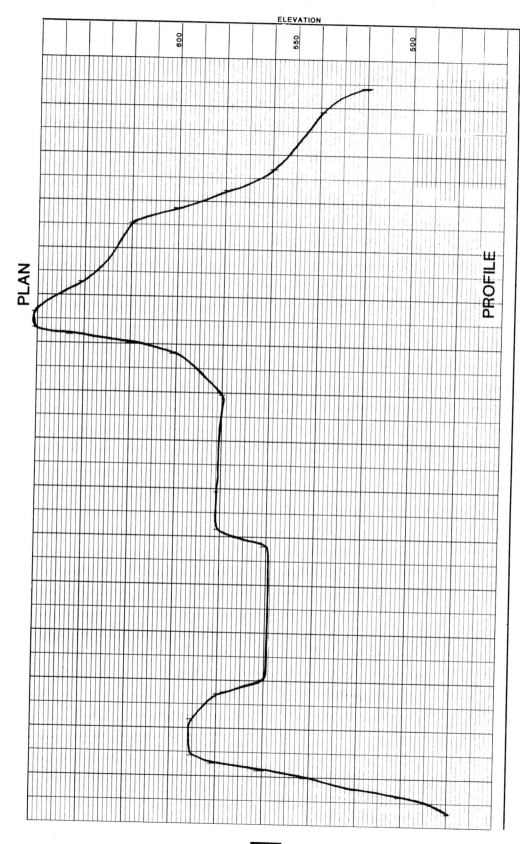

FIG. 7-7 Plotting by contour drawing projection

1. Secure an existing contour drawing, and plot a profile by the projection method.

2. Plot a profile using the following data. Assume that the stations are the center line for a proposed road with a bearing of due East.

Station	Elevation
0	97.6
1	98.2
2	98.9
2 + 65	99.0
3	97.9
4	93.6
TP 1	93.42

Station	Elevation
4 + 55	90.2
+ 65	87.3
+ 75	82.8
5	84.1
5 + 75	88.3
6	90.1

3. Calculate the grade for the following situations:

Elevation	Horizontal Distance
from 890.0 to 897.0	13 feet
from 935.5 to 885.5	27 feet
from 47.2 to 49.9	7 feet
from 134.5 to 177.6	257 feet
from 356.0 to 234.4	375 feet

4. Secure a set of profile leveling field notes and plot the profile, and make the mathematical calculations for elevation checks.

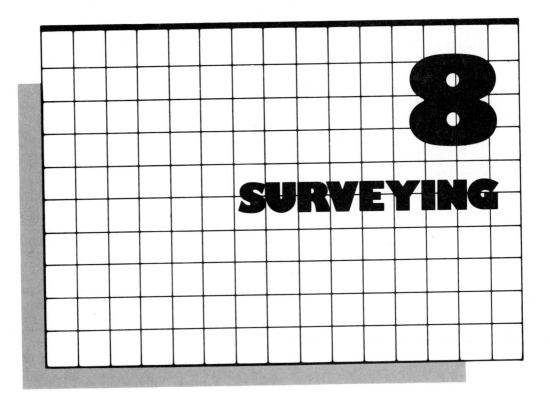

8 SURVEYING

Surveying is the science and the art of determining relative positions of points above, on, or beneath the surface of the earth, or establishing such points.

Mathematical calculations are used in conjunction with the actual measurements of surveying. Distances, directions, locations, areas, elevations, and volumes are determined from data obtained from a survey. Maps, profiles, cross sections, and diagrams are constructed from such information.

The process of surveying is divided into field work (taking the actual measurements) and office work (doing the computing and drawing necessary to the purpose of the survey).

8.1 USE OF SURVEYS

Surveys are divided into three categories: (1) those which establish land boundaries, (2) those which provide information for the construction of public and private works, and (3) those which provide high-precision data for use by federal and state governments. There are no major differences among the three categories as regards survey methods.

The purpose of the earliest surveys was to determine territorial boundaries. Such surveys are still important today.

Except for land surveys, almost all surveys—public and private—help in the conception, design, and execution of engineering works. The surveyor takes measurements and constructs lines and points from the measurements. Every construction project is based upon such data.

The federal government and some state governments conduct surveys over large areas for many years. These surveys have many different purposes. The fixing of state and national boundaries, the charting of coastlines, streams, and lakes, and the precise location of reference points throughout the United States have constituted the principal work accomplished through such surveys. The surveys also collect valuable facts about the earth's magnitude at widely scattered locations, and map portions of mineral deposits in the interior and in older and thickly settled regions.

8.2
THE EARTH, A SPHEROID

The earth is an oblate (flattened at the poles) spheroid of revolution. The length of its polar axis is less than the length of its equatorial axis. In 1866, Frank Clarke determined that the polar axis is 41,710,242 feet in length and the equatorial axis is 41,852,124 feet in length. These lengths have been generally accepted in the United States, and have been used in government land surveys. However, John Hayford's calculations are considered to be more correct than those of Clarke. In 1909 Hayford calculated the length of the polar axis at 41,711,920 feet and the length of the equatorial axis at 41,852,860 feet. The International Geodetic and Geophysical Union adopted figures published by the United States Naval Observatory in 1924. The adopted equatorial axis figure is 41,852,860 feet. The polar axis figure is computed from the equatorial axis figure by assuming that the flattening of the earth is exactly 1:297. Therefore, the polar axis figure is 41,711,940 feet.

According to these calculations, the polar axis is shorter than the equatorial axis by approximately 27 miles. This is a very small amount relative to the earth's size. To illustrate this concept, imagine the earth shrunk to the size of a billiard ball, but having its same shape. To the eye this would appear to be a perfectly spherical surface. Only by exact measurements could the flattening at the poles be detected.

Now imagine that the irregularities of the earth are removed; the total surface of this spheroid is curved. This means that every point on this surface is normal to the plumb line. This surface is termed a *level surface*, and at average sea level, it is called *mean sea level*.

If a plane were passed through the center of the earth, its intersection with the level surface would form a continuous line around the earth. A *level line* would be any part of this line. A *great circle* would be the intersection of the plane with the mean level of the earth. The distance between these two points (the end points of the level line) is the length of the arc of the great circle passing through the points (points A and B in Figure 8-1). This length is always greater than the length of the chord intercepted by this arc. The arc is a level line, and the chord is a mathematically straight line. See Figure 8-1.

The line defined by the intersection of the level surface and plane is called a *meridian* if the plane is passed through the poles of the earth and any other point on the earth's surface. Imagine that two such planes pass through two points on the earth. (See points A and B in Figure 8-2.)

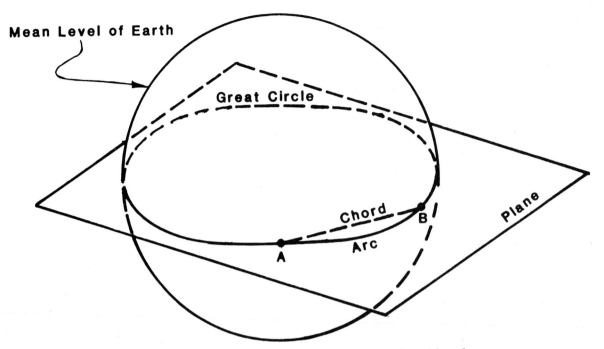

FIG 8-1 Relationship between straight chord and earth's surface

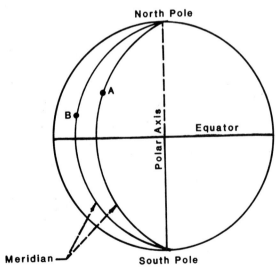

FIG 8-2 Relationship between meridians

The section between the two planes would resemble an orange slice, Figure 8-3. Their meridians are parallel at the equator. They meet above and below the equator at the respective poles. The angle of conver-

gence increases as they approach the poles. The two meridians are parallel only at the equator.

If the orange slice-shaped geometric figure is dissected into pie shapes (Figure 8-3), the lines will be at the center of the sphere. The lines dividing the slice at all points are not parallel. The radial lines may be considered as vertical or plumb lines. Therefore, all the plumb lines converge at the earth's center, but no two lines are parallel. (This is not always true, because of the unequal distribution of the earth's mass. Normals to an oblate spheroid do not always meet at a common point.)

If the mean surface of the earth has three points on it, the three points are the vertices of a triangle. The area in this triangle is a curved surface. Therefore, the lines forming the sides of the triangle are arcs of great circles. This figure is a *spherical triangle*. The dotted lines from the three sides of the triangle represent the plane triangle whose

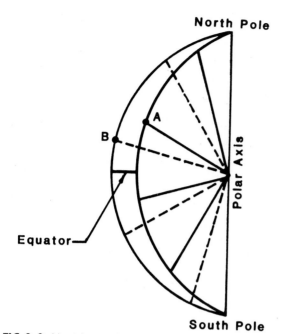

FIG 8-3 Meridian relationship with equator

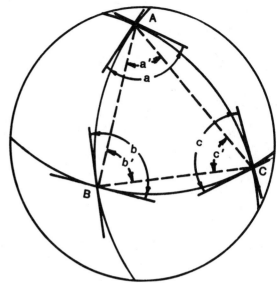

FIG 8-4 Spherical triangle ABC

vertices are points *A, B,* and *C* (Figure 8-4). (The *auxiliary plane triangle* of geodetic work has sides equal in length to the arcs of the corresponding spherical triangle.) Lines are then drawn tangent to the three sides at the triangle's vertices. The measurements of the spherical triangle are greater than the measurements of the triangle of the lines tangent to the spherical triangle. If the points were close together, the difference in the measurements would be small. The difference would be considerable only if the points were farther apart. The same conditions pertain for a figure of any number of sides. Therefore, the *spherical angles* are the angles on the surface of the earth.

We are not concerned with these facts in everyday life, because only a small portion of the earth's surface is observed at a time. A line passing along the surface of the earth directly between two points is thought of as being a straight line. The level surface is thought to be a flat surface. The angles between lines on such surfaces are considered to be plane angles.

The amount of precision required in a survey depends upon whether the survey measures the curved surface of the earth or measures the surface as a plane. It is much less complicated to measure the surface as a plane.

8.3
PLANE SURVEYING

A *plane survey* is a type of survey in which the mean surface of the earth is considered to be a plane. The spheroidal shape is neglected. A level line is considered as mathematically straight with regard to the horizontal distance and direction. Within the limits of the survey, the direction of the plumb line is the same at all points. All angles are considered to be plane angles. Most surveys are of this type.

When the length of an arc 11.5 miles long, lying on the earth's surface, is only 0.05 feet greater than the sustained chord, the shape of the earth must be taken into consideration. This is necessary in surveys of precision covering large areas. The difference between the sum of the angles in a plane triangle and the sum of the angles in a spheroid triangle is only 1 second for a triangle at the earth's surface having an area of 75.5 square miles.

Plane surveys are used for the location and construction of highways, railroads, canals, and local boundaries. (They are not used for establishing state and national boundaries.) The United States' system of subdividing public land also employs the method of plane surveying, which takes into account the shape of the earth in the location where the primary lines of division are drawn.

The operation of determining elevation is usually considered a division of plane surveying. Elevations are made in reference to a spheroid surface, a *spheroid surface* being a tangent at any point on the surface which is normal to the plumb line at that point. The curved surface, usually at mean sea level, is referred to as *datum*. (This is sometimes incorrectly called a *datum plane*.) Calculations to find the elevation of the earth utilize the curvature of the earth. These calculations are usually secured without a special effort on the part of the surveyor.

It is easier to refer to a true plane than to an imaginary spheroid surface. Imagine a true plane, tangent to the surface of the mean sea level at a given point. Suppose that the vertical distance (elevation) of the plane above the surface at a horizontal distance of 10 miles from the point of tangency (represented by mean sea level) is 67 feet. At a distance of 100 miles from the point of tangency, the elevation of the plane is 6670 feet above mean sea level. This demonstrates that the curvature of the earth's surface is a factor that cannot be neglected in obtaining even very rough values of elevation.

8.4 GEODETIC SURVEYING

A *geodetic survey* is a type of survey that takes into account the curvature of the earth. Surveys involving this principle are of high precision and generally extend over large areas. For an area that is not extremely large, perhaps the size of a state, the precision required may be found by assuming the earth is a perfect sphere. If the area is very large, such as an entire country, the true shape of the earth is considered.

Surveys on very large areas are done only by government agencies. The United States Coast and Geodetic Survey and the United States Geological Survey are the only major agencies that conduct such surveys. However, the Great Lakes Survey, the Mississippi River Commission, several boundary commissions, and others have conducted similar surveys. Larger cities like Washington, DC, Baltimore, Cincinnati, and Chicago have conducted the surveys under the assumption that the earth is a perfect sphere.

Geodetic survey data are important because they furnish exact points of reference. Most surveys of lower precision may be tied to these reference points, even though few engineers and surveyors are employed in geodetic work. A system of plane coordinates for each state has been devised. This way all points in the state can be referred to without great error in distance or direction coming from the difference between the reference surface and the actual mean surface of the earth.

8.5 OPERATIONS OF SURVEYING

The many operations of surveying are discussed in this section.

Land Surveys

The *land surveyor* reruns old land lines to determine their length and direction, and reestablishes old or obliterated land lines from recorded lengths and directions along with as much other information as can be secured. The land surveyor also subdivides land into parcels of predetermined shape and size, and sets monuments to preserve the location of such land lines. These monuments are located with respect to permanent landmarks.

The surveyor calculates the area for distance, angles, and directions. This data is drawn onto a land map which shows all the important information found in the survey. A description is then written containing all the surveyed information for deeds.

Topographic Surveys

A *topographic survey* is made to secure field data from which a topographic map can be made to indicate the relief, or elevations and inequalities, of the land surface. The locations of natural and artificial objects are also included in such maps.

The horizontal location of certain points by angular and linear measurements for the skeleton of the survey is called the *horizontal control.* The *vertical control* is the determination of the elevation of control points by the operation of leveling. The topographic survey also determines the horizontal locations and elevations of a sufficient number of ground points to provide the data for the topographic map. The locations of natural and artificial details are required data. As in land surveys, the topographical survey calculates the angles, distances, and elevations needed for the plotting and finishing of the topographic map.

Route Surveys

Another type of survey is the *route survey.* This type of survey is needed for the location and construction of lines of transportation or communication. Such surveys take into account highways, railroads, canals, transmission lines, and pipelines. The preliminary work usually consists of a topographic survey. The location and construction may establish the center line by setting stakes at intervals, running levels to determine the profile of the ground along the center line, plotting such a profile, and fixing grades. By taking cross sections, the volumes of the earthwork are calculated. Drainage areas are measured to lay out structures, such as culverts and bridges, and right-of-way boundaries are located.

Hydrographic Surveys

Hydrographic surveys deal with bodies of water for purposes of navigation, water supply, or busaqueous construction. First a topographic survey is made of the shores and banks. Soundings are taken from the water to determine the depth of the water and the character of the bottom. The soundings are located by angular and linear measurements. The fluctuation of the ocean tide or the change in the level of lakes and rivers is observed along with the discharge of streams.

The shores, banks, the depths of the soundings, and other details are plotted on a hydrographic map.

Actually, hydrographic surveys are undertaken for purposes of drainage and irrigation. However, most of the work is either topographic or route surveying.

Mine Surveys

Mine surveying makes use of the principles of land, topographic, and route surveying, although modifications are made in practice.

Both surface and underground surveys are needed in mine surveying. These establish the surface boundaries of claims for mineral patents and fix reference monuments. In addition, surface shafts, adits, boreholes, railroads, tramways, mills, and other details are located. A topographic survey of the mine is then made, and the results are recorded on a construction map. Underground surveys are also taken to make sure the mine is working efficiently. If underground work is to be done, construction plans are made showing the workings, longitudinal section, and transverse section. The geological plan is then constructed, and the volumes to be removed are calculated.

Cadastral Surveys

Cadastral surveying is the locating of urban and rural property lines. Cadastral surveys provide improvements in detail. These surveys are undertaken mainly for the use of extent, value, ownership, and transfer of land. The term *cadastral surveying* is sometimes applied to the primary control of the United States' public-land surveys.

City Surveys

City surveying is the laying out of lots, city streets, water supply systems, and sewers. There is not much difference between city and cadastral surveying, except in the degree of precision of measurements. The degree of precision of the survey is proportional to the value of the land.

City surveying has come to be associated with a survey of the area in and near a city for the main purposes of fixing monuments, locating property lines and improvements, and determining the configuration and physical features of the land. City surveys have value for a wide variety of purposes. They are particularly valuable in planning city improvements. Like topographic surveys, city surveys establish horizontal and vertical control. A topographic survey and map are first made of the area. Critical points such as street corners with monuments are referred to through a common system of rectangular coordinates. A property map of the area will show the layout and dimensions of properties. This information is used to make a wall map.

Smaller maps show the underground utilities of the city.

Photogrammetric Surveys

Photogrammetric surveying, usually topographic work, is the science of taking measurements by means of photographs. The photographs are taken with specially designed cameras from either airplanes or ground stations. The principle of *perspective* is the minute detail from the photograph transferred to a scaled map.

To accurately establish ground points where space is not available, aerial photogrammetry is used in topographical surveys to ensure that adjustments and projections are correct. Important simplifications and advancements in this technique have made the growth of this method the most rapid. It is extremely accurate, except where the ground is relatively flat, where forest growth is heavy, or where the area is small.

There are three advantages to aerial photogrammetry: the speed with which the work is accomplished, the wealth of detail secured, and its access to locations which are otherwise difficult or impossible to reach. This method is used in general topographic surveys, route surveys, surveys of agricultural areas, and for military purposes. A great deal of the United States has already been photographed. These photographs are available to surveyors and others for a nominal fee from federal, state, and local governmental agencies.

Terrestrial Surveys

Terrestrial surveys, or photographic surveys from ground stations, have been found useful in the small-scale mapping of mountain areas. The photographs, taken from one or more control stations, are used for utilizing the projection of details of the terrain in plan and elevation.

8.6
KINDS OF SURVEYING AND MAPPING

The American Society of Civil Engineers (ASCE) adopted classifications of surveying and mapping in 1959. These are shown in Table 8-1. ASCE adopted the policy of items 1-4. The Surveying and Mapping Division of ASCE maintains technical committees on cartographic surveying, engineering surveying, geodetic surveying, and land surveying. The American Congress on Surveying and Mapping has divisions dealing with property surveys, control surveys, topography, cartography, and instruments. Photogrammetry is within the province of the American Society of Photogrammetry.

8.7
DEFINITIONS

Some commonly used terms in surveying are defined here.

A *level surface* is a curved surface, every element of which is normal to the plumb line. It is parallel with the mean spheroidal surface of the earth, disregarding local deviations of the plumb line to the vertical. The best example of a level surface is a body of water.

A *horizontal plane* is a plane tangent to a level surface. A *horizontal line* is a line tangent to a level surface. In surveying, it is commonly understood that a horizontal line is straight. An angle formed by an intersection of two lines in a horizontal plane is called a *horizontal angle*.

A *vertical line* is a line perpendicular to the plane of the horizon. An example of this is a plumb line. A *vertical plane* is a plane of which a vertical line is an element. An angle between two intersecting lines in a vertical plane is a *vertical angle*. It is understood in surveying that one of these lines is horizontal. A vertical angle is thought to be an angle in a vertical plane between a line to that point and the horizontal plane.

Measured angles are either vertical or horizontal.

In plane surveying, a *horizontal distance* is the distance measured along a level line. From the plumb line through one point to the plumb line through the other point, this distance between two points is commonly understood to be the horizontal distance. The measured distance may be either inclined or horizontal. The inclined distances are usually reduced to equivalent horizontal lengths.

The vertical distance above (or below) some arbitrarily assumed level surface, or datum, is the *elevation* of a point.

Table 8-1
Kinds of Surveying and Mapping

1. Land or Property Surveying (Cadastral)
   a. Property and boundary surveys
   b. Subdivision surveys and plats
   c. Public-lands surveys
   d. Surveys for plans and plats

2. Engineering Surveys for Design and Construction
   a. Design data surveys (including route surveys)
   b. Construction surveys
   c. Mine surveys

3. Geodetic Surveying, Geodetic Engineering, or Geodesy
   a. Control surveys, first- and second-order accuracy
   b. Geodetic astronomy
   c. Gravity surveys, magnetic declination surveys, figure-of-the-earth studies

4. Cartographic Surveying, Cartographic Engineering, or Map and Chart Surveying
   a. Control surveys, third- and fourth-order accuracy
   b. Topographic-planimetric surveys and maps
   c. Hydrographic surveys

5. Aerial Survey Series
   a. Aerial photography
   b. Electrical measurements for distances and position fixes
   c. Airborne magnetometer surveys
   c. Radar-altimeter profiles and elevations

6. Cartography (Not Requiring Original Surveys)
   a. Map design
   b. Compilation derived from existing source data
   c. Map editing
   d. Map reproduction

A *contour* is an imaginary line of constant elevation on the ground surface. A *contour line* is the corresponding line on the map.

The *difference in elevation* is the vertical distance between two points. This is the distance between an imaginary level surface having a high point, and a similar surface having a low point. The process of measuring the distance in elevation is called *leveling*.

The *grade* or *gradiant* is the slope of a line, or the rate of ascent or descent.

8.8 UNITS OF MEASUREMENT

The operations of surveying involve both linear and angular measurements. The *degree, minute,* and *second* are the units of measurement used. A plane angle extending completely around a point equals 360 degrees. One degree is equal to 60 minutes; one minute is equal to 60 seconds. The *grad,* or *grade,* is the angular unit in France. Four hundred (400) grads are equal to 360 degrees. The *mil* is used for military operations. In this measurement 6400 mils are equal to 360 degrees.

In a survey, measurements to the nearest minute are adequate. Angles are determined to the tenth of a second on the more precise surveys.

The common units of linear measurement in English-speaking countries are the *inch, foot,* and *yard.* In surveys done in these countries, most distances are measured in feet, tenths of feet, and hundreds of feet. Surveyor's tapes are usually made to measure accurately in these units. For use in building trades such as construction, the surveyor often finds it necessary to use the foot, the inch, and the eighth of an inch. Most measurements in surveying do not need to be taken closer than hundredths of a foot. Sometimes it is not necessary to measure distances smaller than the nearest foot or even to the nearest 10 feet.

The *rod* and the *Gunter's chain* are units of measurement that were formerly used in land surveying. Subdivisions of the United States' public lands still employ the Gunter's chain as a unit of length. The Gunter's chain is equal to 100 links and is 66 feet long. One mile equals 80 chains. Eighty chains equal 320 rods.

The *meter* is the main unit of measurement. One meter is equal to 39.370 inches or 3.2808 feet or 1.0936 yards. The United States Coast and Geodetic Survey uses the meter as the unit of length.

Mexico and many countries influenced by Spain use the *vara* as the linear unit of measurement. The surveyor must often rerun property lines from old deeds of states formerly belonging to Spain or Mexico because the deeds were written using the vara. Commonly, 1 vara equals 32.993 inches in Mexican measurements, 33 inches in Californian measurements, and 33.5 inches in Texan measurements. There are other variations of the vara.

The *square foot* and *acre* are common units of measurement in the United States. The *square rod* and *square Gunter's chain* were used formerly. One acre equals 10 square Gunter's chains. Ten square Gunter's chains equal 160 square rods. One hundred and sixty square rods equal 43,560 square feet.

The volumetric units of measurements are the *cubic foot* and the *cubic yard.*

Measurement conversions are included in the Appendix.

8.9 PRECISION OF MEASUREMENTS

It is common to think in terms of exact values when dealing with abstract quantities. Students of surveying should appreciate dealing with physical measurements that are correct. Students should also recognize that errors cannot be completely eliminated. The degree of precision of measurements depends on the methods and the instruments used, and on the conditions under which the survey is done.

It is necessary to achieve accuracy in surveying. It is sometimes difficult to achieve exact measurements because of the time and effort needed to conduct the survey. However, it is the duty of the surveyor to perform a high degree of precision for the purpose of the survey. It is important that the surveyor know the sources of the survey and the different kinds of errors that are possible, their effect on field measurements, and the instruments and methods available to minimize errors.

The surveyor must consider the purpose of the survey. The purpose will determine the degree of precision required. The amount of funding available will also have some bearing on precision. (The more precise surveys require more time to ensure accuracy; such surveys are costly.)

If there is an error, its source must be found.

A method must be followed to prevent error. Also, the proper instruments should be used so that the work is organized to reduce the amount of labor required, and the survey results must be checked for accuracy.

8.10
PRINCIPLES USED IN SURVEYING

In plane surveying, the principles are not difficult. A thorough knowledge of geometry and trigonometry is necessary, but less knowledge is required in physics, astronomy, and the theory and methods of adjustment of errors. The knowledge of these last three subjects is necessary for other types of surveying. Geodetic surveying requires an expert knowledge of all of the above subjects.

8.11
PRACTICE OF SURVEYING

The practice of surveying is complex, like other arts that are based upon the sciences.

An individual surveyor's knowledge of the theory of surveying is valuable only if the surveyor is also knowledgeable in the skills of observing as well as field and office procedures. In other words, the surveyor should know the practical phases of surveying as well as the theories involved.

Surveying is one of the first professional subjects studied by engineering students. Even if the student does not plan to become a surveyor, it is good to be trained in the art of observing and computing, the study of errors and their effects and causes, and the practice of mapping. Training in these subjects can contribute to one's success in learning other subjects.

8.12
REQUIREMENTS FOR SURVEYORS

It is important that the surveyor have a thorough knowledge in the theory of surveying and skill in its practice. Employers of surveyors suggest that good character and the ability to work hard aid in the success of a surveyor's career. They also say that initial technical knowledge or skill is not as important as potential abilities. It is important for the student to learn the theories and practice of surveying while developing good work skills and habits.

A surveyor should be reliable, have sound judgment, and possess initiative so as to attack problems with energy and knowledge. It is important to doublecheck all answers by assuming that none are correct unless checked more than once. By doing this, the surveyor will achieve a scientific attitude. This ability to reason logically is important. All projects should be finished thoroughly and in a neat, orderly fashion.

The purpose of this section is to provide the student with a practical knowledge of drawings and related calculations that are connected with plane surveying. It is not necessary to consider the curvature of the earth in plane surveys because the surveys are performed on limited areas. Plane surveys deal with the relative location of points on or near the surface of the earth. Geodetic surveying takes into consideration the shape of the earth.

The surveyor must have a good background in the use and care of drafting instruments and equipment. Greater care is taken in the making of survey drawings than in regular mechanical or architectural drafting. Considerable care and skill in plotting and drafting are required to keep the desired consistent relationship between the field measurements and the completed map.

Scaled dimensions from the drawings are not used frequently, although dimensions are usually noted and scaled in mechanical drawing. In the drawings connected with civil engineering, many dimensions are omitted. It is necessary for the user of the map to rely on scaled distances and angular values that are obtained by measurements from the map.

Survey Drawings

The ordinary drawings connected with plane surveying consist of maps, profiles, and cross sections.

Maps. A *map* is a graphic representation of a certain portion of the earth's surface which uses lines and symbols. There will always be some distortion in a finished map because the earth's surface is curved. This surface must be recorded on the plane on a sheet of paper. The distortion cannot be measured by ordinary means since the area measured in plane surveying is small. The drawings are, therefore, made as if the area surveyed and mapped were flat. All the measurements taken in plane surveying are horizontal measurements or the measurements are reduced to the equivalent horizontal measurements before being used in map plotting.

Profiles. A *profile* is a drawing showing a vertical section along a certain survey line. Profiles are made by plotting the elevations of points located at rather short measured distances along a certain located line. Examples are the center line of a highway, a railroad, and a canal. An elevation is the vertical distance of a point above or below a fixed reference plane. Mean sea level is the reference plane that is most commonly used in plane surveying.

Profiles are along and in the direction of the line surveyed, and are plotted as a vertical section.

The scales used are determined from the roughness of the earth's surface along the line surveyed. Vertical and horizontal scales are necessary. The vertical and horizontal scales are not usually the same. Profiles are usually plotted on square, cross-section paper, or plan-profile paper.

Cross Sections. A *cross section* is a drawing showing a vertical section usually at right angles to the survey line. Information for plotting cross sections is found by running elevations along lines usually located at right angles to the survey line. Readings for elevations are taken on the cross line at places where there is a change in the slope of the terrain. Elevations can also be found from different readings, by recording the elevations only where the readings show a selected contour value. The cross lines to the survey line are spaced or located at changes in direction of the survey line or at whole station numbers. These cross lines are also placed at any other location along the line where additional information is needed. The particular topography may also dictate that the information is needed. The cross lines provide information for location studies, earthwork calculations, drainage studies, and locations. They give primary information to determine the shape and configuration of the ground surface on each side of the survey line.

Traverses

In surveying, a *traverse* is a succession of points connected by straight lines, the directions of the lines and the distances between points having been measured. There are two kinds of traverses. A traverse that begins at a certain point and returns to the point of beginning is called a *closed traverse*. This traverse is used in surveys that deal with a tract or parcel of land. An *open* or *continuous traverse* is one that does not return to the point of beginning.

Station numbers are usually used to verify the distances along an open traverse. A *station* is a unit of 100 feet. This is referred to as a *full station*. A *plus station* is a fractional part of a full station. The beginning or starting point in an open traverse is called *station 0+00*. Open traverses are usually numbered in stations continuously from the beginning point to the end of the traverse. A point at a distance of 782.52 feet from the beginning of an open traverse would be numbered 7+82.52. The *7* is the number of the full station, and the *82.52* is the fractional part of a full station.

Stakes or markers are set out at regular intervals on the line during the operation of running a traverse. They are usually set out at intervals of 100 feet. The distance between stakes is sometimes reduced to 50 feet, 25 feet, or less. Each line on a traverse is known as a *course*.

Examples of open traverses are the surveys for highways, railways, canals and pipelines.

Distances

The distances measured in the field are usually measured with a 100-foot-long steel tape. The process of measuring distances is called *taping* or *chaining*. Distances are almost always measured in feet. The Gunter's chain was the most common unit of measurement at one time. It was used, by law, in the original surveys authorized by the United States government.

Two types of measurement of angles and distances are encountered. One type is *observed quantities*, which are actually observed or measured in the field. The other is *calculated quantities*, which are calculated from measured quantities.

To find the actual distance between any two stations, subtract the smaller station number from the larger station number.

The actual distances between changes in directions of the line are recorded in feet instead of in station numbers in closed traverses.

All measurements taken during the survey are recorded in a notebook, which is referred to as a *field notebook* or *field book*. Drafters use the notes and figures in the field book when making the drawings and maps required. Erasures are not allowed in a field book. A line is drawn neatly through any incorrect figure, and the correct figure is then wirtten in its place.

Kinds of Maps

There are many types of maps, such as topographic, geographic, planimetric, real estate, road, and land survey maps. A topographic map shows the shape and configuration of the ground surface. If the relief of the ground surface is not shown it is a planimetric map. Geographic maps show the location of communities, roads, railroads, political subdivisions, and other important information.

Scales

The drawings related to surveying are drawn to scale. A certain distance on the map represents a proportional distance on the ground. The *scale* of the map is the ratio of a distance on the map to the corresponding distance on the ground. The scale may be proportioned for 1 inch on the map to equal or represent a certain number of feet on the ground. The scale may also be represented by a fraction showing the ratio of the scale to the actual ground measurement. For example, *1/1000* means that 1 unit of measurement on the map represents 1000 units of measurement on the ground. These are both scaled to the same unit of measurement, whether it be miles, yards, feet, or inches. The graphical scale is another type of scale. This scale is drawn on the map to represent the actual distance. The advantage of the graphical scale is that any distortion due to shrinkage or expansion of the paper also distorts the scale equally. The scale or ratio will then remain true. A noted scale is one that specifically states that 1 inch equals 1000 feet. This means that every 1 inch on the map equals 1000 feet on the ground.

North

In order for the user of the map to orient the direction on the ground, North is pointed out specifically. North is indicated by an arrow.

Titles

Each map should be given a title. The name or subject of the map, the authority or who made the map, the direction of the meridian, the scale, and other information needed to use the map properly should be given.

Symbols

Objects are represented on a map by certain signs and symbols. Many of these are conventional—so widely used that they are recognized easily.

Lettering

Lettering on a map is of great importance. A perfectly neat drawing can be ruined by poor lettering. Good lettering ability can be acquired only by study and frequent practice. It is assumed that most civil engineering students have learned the principles of lettering in basic drawing courses.

Notes and Legends

Notes and legends are used on a map for explanatory purposes to avoid any questions by the map's users. The notes should be brief but clearly understood. Notations at a convenient place on the map should explain anything out of the ordinary, and legends should list any unfamiliar or uncommon signs or symbols. A key to any widely accepted symbols should also be shown on the map. An acceptable place for notes and legends is near the title block, where they can be seen easily.

8.14 SUMMARY

Surveying is the science and the art of determining relative positions above, on, or beneath the surface of the earth, or establishing such points. Distances, directions, locations, areas, elevations, and volumes are found from data obtained from the survey.

Surveys determine territorial boundaries and are basic to engineering and construction projects. The federal and state governments use surveys which provide highly precise data.

The earth is an irregular sphere. Certain measurements are taken from determining the spheroid at many different angles.

Plane surveying is required of the earth's surface for the location and construction of highways, railroads, canals, and boundaries. Geodetic surveying uses the shape of the earth for large-scale surveys of states and countries for use by governmental agencies. Many different types of surveying are needed to accurately represent data.

In order to understand the process of surveying, a knowledge of the basic surveying definitions and measurement units is required.

In applying surveying's principles, one must be knowledgeable in trigonometry, geometry, physics, astronomy, and the methods of adjustment of errors.

A map represents all the data figured and calculated during a particular survey. Vertical sections are shown by a cross section. A traverse is a succession of points connected by straight lines, with the directions of and distances between points measured.

Distances, which are recorded in a field book, are needed on any map. There are many types of maps used for surveying. On all maps, North is depicted in order to give direction. Titles, symbols, lettering, scaling, and explanatory notes are also needed to complete a proper survey.

Azimuth
Cadastral Survey
Equatorial Axis
Geodetic Survey
Great Circle
Horizontal Distance
Horizontal Line
Horizontal Plane
Hydrographic Survey

Latitude
Level Line
Level Surface
Leveling
Longitude
Meridian
Mine Survey
Oblate
Photogrammetric Survey

Plane Survey
Polar Axis
Spheroid
Terrestrial Survey
Topographic Survey
Vertical Angle
Vertical Line
Vertical Plane

REVIEW

1. What is surveying?
2. Surveys are divided into what three categories?
3. Name three types of plane surveys.
4. What is aerial photogrammetry?
5. What is a vertical plane?
6. What is a Gunter's chain?
7. Name some characteristics of a good survey.

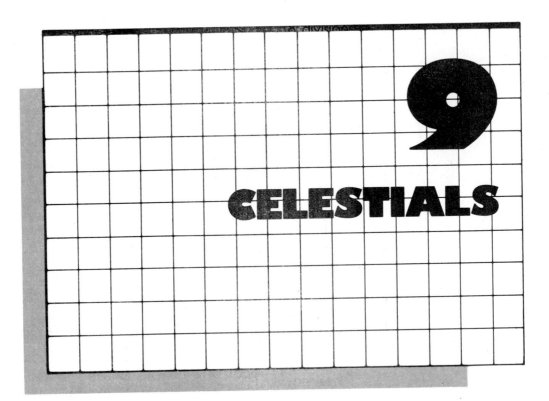

9 CELESTIALS

The use of celestial observations, as discussed in this book, is important for collecting topographic base data. Surveyors make use of the sun, moon, and stars to accurately locate and define the dimensional characteristics of land features and reference points. These observations are based upon the established principles of astronomy.

The process of applying astronomical principles to surveying techniques is highly sophisticated and mathematically oriented. The purpose of this chapter is to briefly describe the elements and procedures used in celestial observations that apply to data collection for map and topographic drafting. Therefore, this discussion does not cover the specific mathematical calculations used by surveyors.

Engineers and surveyors need basic knowledge in astronomy to determine data specifications such as latitude, time, longitude, and azimuth. For a point on the earth's surface, these specifications can be made in relationship to a given celestial body (e. g., the sun, moon, or a star). The specifications are expressed in terms of calculated parameters.

When making celestial calculations, the various heavenly bodies are considered to be located on the surface of a sphere that has a limitless radius. The center of that sphere is assumed to be the center of the earth.

The study and observation of celestials require readings and measurements of horizontal and vertical angles. The angles and triangles formed aid in the solution of surveying problems that incorporate celestials. Unlike a flat surface or plane, these angles and triangles are based upon spherical characteristics; that is to say, the sides of a triangle are curved. The triangle consists of a *zenith*, which is the celestial North pole, and two other reference points, usually another heavenly body and a point on the earth's surface.

A *sextant* is a hand instrument used for measuring horizontal and vertical angles in relationship to celestial bodies. The sextant can be used to find the exact location of a point on the earth's surface. The instrument usually employed by surveyors is the *transit*. The transit is similar to a telescope. It is used for field surveying observations because of its ability to rotate about a full vertical axis while making visual observations.

Observations are calculated by referring to a variety of tables. Examples of these are trigonometric tables, ephemeris tables (tables of celestial bodies at regular time intervals), and tables for Polaris and solar calculations. Surveying teams also use the *American Nautical Almanac*. The *Almanac* is published each year and gives the positions of major celestial bodies at different times of the year.

	9.2 TIME

In addition to angular notations and calculations, time is noted carefully. When any observations and calculations are made, the time must also be given. The reason for this is that every celestial body occupies a different location in the sky at different times. If the time were not given, location calculations could be off anywhere from several feet to hundreds of miles. Thus, without noting the time, celestial calculations in surveying would be useless.

Three types of time are considered in astronomical observations. These are *apparent solar time* (or true solar time), *sidereal time*, and *mean solar time*.

Apparent solar time is determined by the angle (hour angle) of the sun. This angle is measured west of the meridian. An example of an apparent solar time measurement is time indicated on a sundial. Because of its nonuniformity of measurement, apparent solar time is no longer used.

Sidereal time is calculated directly from celestial observations. Sidereal time is based upon the *sidereal day* (which is equal to 23 hours, 56 minutes, 4.09 seconds of mean solar time). The sidereal day is defined as the interval between two transits, in succession, of the vernal equinox (sun crossing the equator). In surveying calculations, it is the hour angle of a point among celestial points. Hence, sidereal time is the hour angle of the March equinox (sun crossing the equator on March 21) at a given point.

Mean solar time is calculated by a theoretical factor known as the *mean sun*. The mean sun day is based upon the principle that the earth will make one complete revolution every 24 hours (hence the 24-hour day). Use of this kind of time assumes that the sun travels at a regular and constant rate along the equator, that it has a consistent relationship to the stars, and that its movement equals the average (mean) speed of the true sun.

A *solar year* contains 365.2422 solar days. In a 1-year period of time, the earth will always face the sun one time less than it does the dark and vertical equinox. This occurs because the earth rotates on its axis each day and orbits the sun only once a year, so the chances of rotating toward (or facing) the sun are less.

9.3 TIME SPECIFICATION

Time at a given location can be specified in one of two ways. *Standard time* is used for greater areas of land. *Local time* specifies time for a given location regardless of its meridian.

Standard Time

Standard time is the mean solar time given to land areas that are between meridians 15° from each other. The measurement is based upon a point of origin at 0° longitude. The 0° longitude passes through an English town called Greenwich, hence the expression *Greenwich Mean Time* (GMT).

The 15° hour spacing is easy to understand. It was calculated by dividing 360° (angle of circle) by 24 (hours in a day), which equals 15°. Therefore, each 15° meridian represents 1 hour of standard time.

Eastern Standard Time (EST) begins at the 75th meridian west of Greenwich; *Central Standard Time* (CST) starts at the 90th meridian. *Mountain Standard Time* (MST) starts at the 105th meridian, and *Pacific Standard Time* (PST) begins at the 120th meridian. If one were starting at Greenwich Mean Time at 6:00 pm, the time at Eastern Standard Time would be 11:00 pm (meridian 75 divided by 15° equals 5 hours; 5 hours plus 6:00 pm equals 11:00 pm).

Local Time

Local time is based upon the exact meridian of a point. *Local Civil Time* (LCT), in relation to mean solar time, is an hour angle measured from the local meridian plus 12 hours. *Local sidereal time* is the hour angle of the vernal equinox from the observer's meridian.

At a meridian in longitude 121° 33' 20", standard time would be calculated upon the 120th meridian. This equals 8 hours plus GMT. If the time is 10:00 am PST, local time is earlier by an amount corresponding to 1° 33' 20". Thus the local time at the given meridian is not 10:00 am PST, but 9:53:46.7 am. Local time is specified in terms of Local Civil Time, and is noted as LCT.

The LCT at longitude 72° 03' 14" for the standard mean time of 10:00 am is calculated by adding the standard time to the amount corresponding to 90° minus (72° 03' 14"). The time interval for this difference is equal to 1 hour, 11 minutes, 47.1 seconds. Hence, the LCT is 1:11:47.1 pm. This time conversion is based upon the data in Table 9-1.

It should be noted that the 10:00 am standard time used in our example serves as the reference time for our calculations. Thus, the time is designated as *pm* (*post meridiem*), rather than *am* (*ante meridiem*).

9.4 RADIO TIME

The United States Naval Observatory sends out signals periodically during the day by radio wave. The signals can be re-

Table 9-1
Time Conversion Data

360° = 24 hours		24 hours = 360°	
1° = 4 minutes		1 hour = 15°	
1' = 4 seconds		1 minute = 15'	
1" = 0.067 seconds		1 second = 15"	

ceived on shortwave radio receivers. These same signals are sent by broadcasting stations at noon and at 10 pm EST. The signals are accurate to within 0.1 second. If more accuracy is required, the Naval Observatory will provide correction factors and data.

The Bureau of Standards in Washington, DC, also sends out time signals. These signals are broadcasted through radio station WWV on frequencies 2.5, 5, 10, and 15 megahertz. Superimposed on a 440-hertz (cycles per second) hum, the signals are sent out continuously and resemble the ticks of a clock for every 1 second of time. There is no signal on the 59th second of every minute. On every 5-minute multiple, the tone is discontinued and restarted 1 minute later. On the hour and on the half-hour, there is a voice announcement.

9.5
TRANSIT OBSERVATIONS

Transits are sighting instruments used for obtaining various types of measurements. In addition to traditional land measurements, transits are used for celestial observations and calculations. In this section the two basic types of transit observations are presented: sun and star observations.

Sun Observations

Sun observations are made to determine the sun's altitude and azimuth. This is accomplished by measuring an arc of the horizon, clockwise between a fixed point and the vertical, established through the center of a given object. The *vertical* is a circle on a transit that is used to determine altitude, while the *horizontal* circle shows the azimuth. (Vertical pertains to the alignment of the sun on the vertical hair line through the transit, while horizontal refers to the alignment of the sun on the horizontal hair line.)

It is difficult to place the cross hair lines of the transit directly on the sun (due to the sun's glare). Therefore, the cross hairs are made to be tangent to the edges of the sun. The horizontal and vertical cross hairs and stadia hairs are bisected by a pair of vertical marks with the same spacing as the stadia hairs. Inside the two tick marks and stadia hairs is the *solar circle,* with an angle radius of 00° 15' 45". This measurement is slightly smaller than the semidiameter of the sun, which is 00° 16'. Thus, the sun's image can be centered on the circle of the *reticule* (the eyepiece of the transit) with great accuracy. See Figure 9-1.

The sun can be viewed directly through a telescope if a dark filter is attached to the eyepiece. When a filter is not used, one of

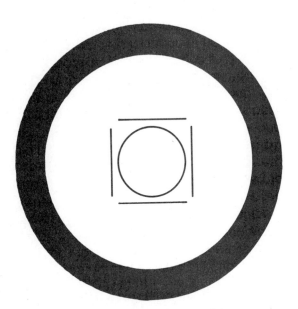

FIG 9-1 How the sun's image would appear through transit reticule

two methods can be used for sighting: the *center-tangent* (or *disappearing-segment*) *method*, and the *quadrant-tangent method*. In both methods a piece of paper is held behind the eyepiece of the transit. See Figures 9-2 and 9-3.

The center-tangent method observes the trailing edge or limb of the sun. The tele-scope is pointed to the sun so that the light will pass through it and show as an image on a white card held 3 to 4 inches behind the eyepiece. The telescope is then ro-tated, both in altitude and azimuth, until the sun's rays cast a circular shadow on the card. This shadow will be upside-down. Thus, the lower edge of the shadow will actually be the top of the sun, and the up-per edge will be the lower edge or limb of the sun. Also, the right and left sides will be reversed. The cross hairs are then focused until they appear clearly on the card, mak-ing the sun's shadow sharp and in focus.

To take azimuth readings on the transit, the vertical-motion tangent screw is turned so that the horizontal cross hair will bisect the sun's image. This measurement is taken in the morning to determine the tangent of the lower limb of the sun. After measuring this point, the cross hair will remain station-ary. The small shadow segment will be-come smaller as the image is moved downward and to the right. The surveyor will then adjust the upper horizontal-motion tangent screw to move the vertical cross hairs. This will cause the cross hair to bisect the disappearing segment. As soon

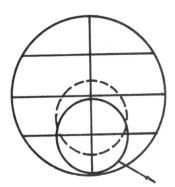

Direction of transit movement

(A) Horizontal crosshair stationary for horizontal reading

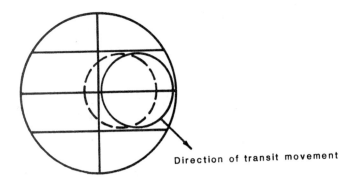

Direction of transit movement

(B) Vertical crosshair stationary for vertical reading

FIG 9-2 Image of sun as seen on card held behind eyepiece of an erecting telescope during morning observation

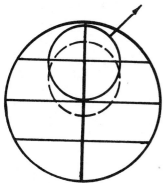

Direction of transit movement

(A) Horizontal crosshair stationary for horizontal reading

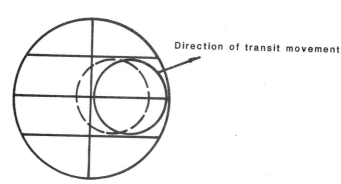

Direction of transit movement

(B) Vertical crosshair stationary for vertical reading

FIG 9-3 Image of sun as seen on card held behind eyepiece of an erecting telescope during afternoon observation

as the edge of the sun's image is tangent to the horizontal cross hair, the surveyor will stop adjusting the screw. At this time, the vertical cross hair will pass through the center of the image, making the horizontal-circle reading correct, and obtaining the aximuth. See Figure 9-2A.

The horizontal cross hair will be tangent to the lower limb during the azimuth reading. The vertical angle must then be increased by 0.5 the angle sustained by the sun. The measurement is the sun's semidiameter (00° 16'). The value of this reading will vary for different times of the year. To obtain the correct value, one must refer to the data provided in the ephemeris. There is no need to add the semidiameter to the vertical angle, as is needed for the horizontal angle. The vertical angle is corrected for refraction, parallax, and index error. See Figure 9-2B.

When a reading is to be taken for the horizontal cross hairs in the afternoon, the upper limb is used. The horizontal cross hairs are first set as in the morning's observations. After the corrections for index error, parallax, and refraction have been made, the semidiameter must be subtracted from the vertical reading. See Figure 9-3A.

Altitude readings are made by setting the vertical cross hair to bisect the sun's image. This reading is taken to the eastern limb of the sun, and can be accomplished in the morning or evening. The horizontal hairs are then moved to bisect the image until tangency is achieved. At this time, the cross hair should cut the center of the image to make the vertical-circle reading (altitude). Correction is then required for index error, parallax, and refraction. See Figure 9-3B.

After each reading is completed and the values are corrected, the time is noted for each observation, to the nearest second. The procedures are then repeated to obtain several azimuth and altitude readings. These readings should be charted according to the time of the observations. Each plot should be a straight line, or close to a straight line. Those values that are not close to the straight line must be rejected before the total computation is made.

The quadrant-tangent method for observing the sun is similar in procedure to that of the center-tangent method. In the quadrant-tangent method, the vertical cross hair is aligned with the sun's western or leading limb by adjusting the upper horizontal-motion tangent screw. When the screw is moved, the vertical motion of the transit will not be disturbed. Therefore, the horizontal cross hair will remain at a constant altitude. Furthermore, the lower limb, which appears as the upper edge of the image, will be tangent to the horizontal cross hair. The exact time is recorded when tangency occurs.

The next step requires that the horizontal cross hair be made tangent with the sun's upper limb (appearing on the card as the lower edge of the image) by adjusting the vertical slow-motion screw. The transit is stopped and the time recorded as soon as the eastern limb becomes tangent to the vertical cross hair. If the reading is made in the afternoon, the sun's upper limb will be made tangent first, followed by the eastern limb.

Surveyors observe tangency at two points of the sun's image when using the quadrant-tangent method, and one point when using the center-tangent method. Because of the two tangency points, greater care and time must be taken to ensure accurate readings and observations.

To eliminate any incorrect readings, several observations are made and averaged. Solar observations are never finalized until all corrections have been made to the circle readings and computed angles have been plotted as a function of time. Plottings can identify mistakes in less time than is required for mathematical calculations. It is frequently presumed that observations are correct and the readings averaged accurately; therefore, the only possible way to identify error is by plotting these observations.

Roelof's Solar Prism is a device which produces four overlapping images of the sun to give a solar disc pattern. This is accomplished by fitting the prism onto the telescope of a transit. By looking through the transit, the four overlapping circles form a diamond-shaped area that can be dissected by both horizontal and vertical hair lines with exceptional accuracy. The location of the exact center of the solar disc is possible because of the symmetry of the pattern. This also eliminates the need for semidiameter and horizontal-circle corrections. Corrections, however, must still be made for vertical-circle observations.

Star Observations

A star, when viewed through a transit, appears as a point of light. This characteristic makes it extremely easy to align with the vertical and horizontal cross hair lines.

Because star observations are made at night, and the star emits a small amount of light, the black cross hair lines are impossible to see without additional light. A technique commonly used in the field to supply additional light is to shine the light from a flashlight at an angle across the lens of the transit. The light source should be at an angle which will make the cross hairs visible but not blot out the image of the star. Many transits come equipped with a rheostat that regulates the internal illumination of light.

The ground stations must also be illuminated if a star is to be sighted for the azimuth of a line, or if a backsight has been made on the ground station during the observation. When the station is being observed, any suitable light source (such as a flashlight) can be employed. The light source is held behind the mark for easy observation.

For convenience in setting up the transit for star sightings, preparation should be made in either daylight or twilight hours. The twilight hours are considered the most desirable time for the surveyor to make observations, because there is still some natural light in which to work, and star sightings can be made. In some situations, however, the twilight hours provide too much light to view many stars.

9.6
TIME BY TRANSIT OBSERVATIONS

The primary purpose of celestial observation is to establish the azimuth of a line. It is therefore important to note the exact time of field observations, since location and time are closely related in this type of field surveying.

Field observations are conducted for time when the standard time is unknown, or if greater accuracy is required. Accuracy can be increased by noting the exact time at the point of observation. Most celestial observations conducted for determining time are made to increase accuracy, since the standard time of most locations is known by the survey crew.

Time can be specified by transit observations by using one of two techniques. One technique is to determine the time that the sun or star crosses the meridian of the observation point. The second technique is to determine the altitude of the sun or star.

Sun Observations

Time for the point of observation may be determined by noting when the sun's center crosses the meridian. The longitude of that point is then determined by a detailed and accurate map reference.

This process is accomplished by first setting the telescope of the transit on the point of observation and its meridian. Next, the sun is tracked until a small part of the sun is bisected by the vertical cross hair. The exact time of this crossing is recorded, and is referred to as *Local Apparent Noon.*

The Greenwich apparent time is then found by adding the longitude in time units to the local apparent time of 12 hours. The equation time that is obtained from the ephemeris is subtracted algebraically to obtain the Greenwich civil time of the observation.

Time can also be determined by measuring the altitude of the sun's center at any instant. The altitude is usually measured to either the upper or lower limb via the transit telescope. To increase the accuracy of these readings, several altitude measures are taken in succession. After corrections for index error, parallax, refraction, and semidiameter are made, the time is computed by using various algebraic formulae.

Star Observations

Time can be determined by star observation by noting when a star crosses the meridian of a point. Unlike the sun, stars' apparent motions tend to be more rapid near the equator. Therefore, the stars chosen for these observations are relatively closer to this point.

Several stars are usually observed for time calculations. The stars are identified by computing their times of transit and location in the sky. The colatitude of the observation point plus the declination of the star will equal the altitude. The altitude is usually kept below 55° to 60°, unless the transit is equipped with a prismatic eye, for greater accuracy and ease of computation.

The formula used for calculating the altitude of the sun is also used to calculate the altitude of a star. At least two stars should be used for altitude specifications. These stars should be directly east and west of the observation point, with the closest star being at least 30° above the horizon.

When measuring the star's altitude, no corrections are necessary. Instrumental and refraction correction errors are largely eliminated from the results because only one star is observed in opposite directions. Furthermore, with half the telescope normal and half inverted, errors are minimized when several altitudes are measured in succession.

9.7
LATITUDE

Latitudes are used to compute time or the azimuth of a line, and are therefore an important specification. To obtain the latitude of a point, one can use an accurate map and scale the latitude through that point. This procedure is usually accurate enough for most topographic and map drawings. However, in situations where greater accuracy is required, other procedures can be used. The procedures include determining latitude by the sun's altitude at noon, by observation on circumpolar stars, and by Polaris at an hour angle.

Latitude by Sun's Altitude at Noon

The latitude may be obtained by measuring the altitude of the sun's lower or upper limb at the time it crosses a given meridian.

The altitude is measured when the sun reaches the highest point in its path for a meridian. For any error in the observed angle corrections are made for factors such as refraction, parallax, and the semidiameter of the sun. The semidiameter of the sun and its declination at the time of apparent noon are found in the ephermeris. It should be noted that north declination is a positive value, while south declination is negative.

Latitude by Circumpolar Stars

A *circumpolar star* is a star that is always visible above the horizon line, or that is found in the vicinity of a terrestrial pole (*e. g.*, the North Star). The altitude of a pole is also a measure of the latitude of the point of observation. When a star is on a meridian, its latitude can be obtained by measuring the vertical angle to a circumpolar star. *Polaris* (*Ursae Minoris*), the North Star, is the star most frequently used by engineers and surveyors. This star is close to the pole and can be easily identified.

The observation is made when the star reaches its highest or lowest point. The highest and lowest points are known as upper and lower culmination, respectively. *Lower culmination* is when the star is on the lower branch of the meridian. *Upper culmination* is when the star is on the upper branch of the meridian. The time of culmination can be found in special tables in the ephemeris.

Latitude from Polaris at Any Hour Angle

Both the upper and lower culminations of Polaris may occur during daylight hours at certain times of the summer. If the exact time, within a few minutes, is known the altitude can be observed whenever the star is visible, and the latitude can be computed without serious error.

There will be less error due to errors in time if the star nearest to culmination is used. Four hour angles of 3, 9, 15, and 21 hours with an error of 1 minute in time will cause an error of 00° 12′ in the computed latitude. Four hour angles of 6 and 8 hours with an error of 1 minute in time will cause an error of approximately 00° 17′ in the computed latitude.

9.8
LONGITUDE

The longitude of a point can be obtained by scaling its location on an accurate map, or by calculating the difference between the longitude of the known point and an unknown point. Longitude can also be calculated for a known point by difference in local time. This can be determined by using the Naval Observatory's radio time signals as the reference point and time (for the 75th meridian).

9.9
AZIMUTH

Azimuths are a standard system of horizontal angle measurement expressed in terms of the relationship between a line and a reference meridian. As is described in Chapter 6, the azimuth of a line can vary from 0° to 360°, with the reference point being North or, sometimes, South.

The azimuth of a line can be determined by using several celestial observation techniques. These include determining azimuth from observation on Polaris at elongation, Polaris at any hour angle, or by solar observations, solar attachment, and equal angles.

Observation on Polaris at Elongation

Solar observations and observations on circumpolar stars are the two most common methods used to obtain the azimuth of a line. The easiest method is to view Polaris at its most western point, called *western elongation,* or at its most eastern point, called *eastern elongation.*

The azimuth of Polaris at elongation can be found from the ephemeris, or can be calculated by using various equations. The star is followed with the vertical cross hair and the exact time is noted. The star is elongated when it appears to move vertically on the cross hair. The horizontal angle is then measured to a fixed point, creating a reference line. To obtain the required azimuth of a line when Polaris is at eastern elongation, the azimuth is added to the clockwise measured angle. If Polaris is at western elongation, the azimuth is subtracted from the clockwise measured angle.

Observations on Polaris at Any Hour Angle

Observations on Polaris can be made whenever the star is visible if the exact time is known. The azimuth and latitude can be determined during the same observation when the altitude is measured. Trigonometric equations are used for azimuth calculations. The effect of errors in time, on the computed azimuth, will be negligible when the altitude is measured. For instance, when the star is near culmination, the error of 5 minutes in time in a latitude of about 40° will produce an error of about 00° 02′ in the azimuth.

Solar Observations

The azimuth of a line can be determined by measuring the horizontal angle from a line to the sun's center, and the altitude of the sun's center, if the exact time is known within 1 or 2 minutes. Four measurements taken with the telescope direct and reversed will give reliable results to within 00° 00' 15" when a 30-second transit is used. The eight measurements, however, must be taken in quick succession. The measured horizontal and vertical angles are corrected to give the true horizontal and vertical angles to the sun's center. A plot is then prepared to show the correct horizontal and vertical angles as a function of time. The eight measurements are plotted for an afternoon observation. The horizontal angle of plots is shown as a straight line, and the vertical angle is represented by two separate and parallel straight lines. The two parallel lines represent direct and indirect observations.

At times, the plotted results for the observation validity will be questioned. To check on the accuracy of the observation, the vertical angle will be determined by expressing the altitude in terms of the sun's hour angle. Hence, the sun's altitude will be a function of time.

The best results are obtained when the local time is between 8 am and 10 am or 2 pm and 4 pm. To obtain the azimuth of the line, the horizontal angle (azimuth) of the sun is used with the horizontal angle between the backsight, and the position of the sun is represented by the mean of the corrected angles.

Solar Attachment

The *solar attachment* is a device that is fastened to the transit. It can rapidly and accurately determine the direction of the meridian. Several types of solar attachments are used, but they all operate on the same principles.

The solar attachment often consists of an auxiliary telescope. It is equipped with a bubble and has two motions at right angles to each other. The attachment's vertical axis lies in the meridian and points to the North pole when the line of sight is directed at the celestial equator.

The main telescope is then elevated through a vertical angle. The angle between the planes of the horizontal cross hairs of the two telescopes is made equal to the declination of the sun. When the smaller telescope is directed toward the sun, the main telescope will be in the meridian. The two telescopes are brought into the same plane by directing the lines of sight to the same distant object, setting off the declination angle. The declination angle is then set on the vertical circle.

The auxiliary telescope is revolved to find the position on its horizontal axis that will bring the bubble to the center of the tubes. The angle between the two telescopic lines of sight will be the declination angle. The main telescope is then clamped into position after being elevated to the altitude of the observation point's colatitude. Next, the telescopes are turned horizontally until the small telescope is directed toward the sun. At that point, the main telescope will be in the meridian. From this the azimuth can be calculated.

Equal Altitudes

Another technique for determining azimuth involves using two equal altitudes. The two altitude readings are taken from the east and west ends of the meridian so that they will be the same distance from the observation point. The celestial body under observation will have a symmetrical path, with respect to the meridian, when the declination of the sun is assumed to be constant. Therefore, the meridian will be midway between any two positions of the sun which are at equal distances above the horizon.

Use of this procedure requires that a horizontal angle be measured between a fixed line of reference and the sun's center at any convenient morning altitude. At the same time, the horizontal cross hair should be located and readings taken of the horizontal and vertical circles. The vertical angle is then increased by the diameter of the sun for the afternoon observation. The sun is tracked until the horizontal cross hair is tangent to the sun's upper limb. The altitude of the sun will then be exactly the same as its altitude in the morning.

The mean, or mid, distance between the two positions of the sun is not the exact south point because the sun's declination is not considered constant. A special algebraic formula is used to determine correction for change in declination. The variation per day of the declination is positive if the sun is moving north. The mean of the two positions of the sun lie to the west of the south point. When the correction is added to 180°, the result is the azimuth of the sun at the middle position when measured from the north.

A series of observations on different stars can be made at night. Since any change in the declination of a star is negligible, no correction is necessary.

9.10
ORDER OF FIELD OBSERVATIONS

Surveying teams that are required to make celestial observations usually follow a sequential order. The order of making observations is time, latitude, and longitude.

For most azimuth observations, time and latitude should be known to great accuracy. This will permit the use of solar observations and the observations of Polaris at any hour angle. Time can be determined from a good watch and checked by radio time signals or by any Western Union Telegraph clock. Latitude can be scaled with a great amount of accuracy from existing maps.

When no preliminary data (field information) is available, an approximate latitude can be found by measuring the maximum altitude of the sun. It is also possible to find local apparent time in the same manner. The maximum altitude will occur at the instant of Local Apparent Noon. A more exact latitude can be found by measuring the maximum or minimum altitude of Polaris or any other convenient star.

Observation on Polaris, at elongation, will make possible a very close approximation of the observed latitude to the true meridian. Three methods can be used to find local time. These are by observation of the sun at local apparent time, by use of a previously located meridian, or by measuring the altitude of the sun.

Results will be easier to obtain by repeating the observation and by using the data from each preceding observation. The location of time, the local time, and the latitude can then be determined with greater accuracy.

The easiest way to determine the longitude is to compare the local mean time with the standard time. Frequently, time is sent out by radio transmitter. When this method cannot be used, direct observations must be made and entered into algebraic and trigonometric equations. There are various astronomical references that contain methods for approximating longitude on the sun, stars, and moon.

The use of celestials is an important procedure for the collection of topographic and other map data. A highly complicated and mathematically oriented field, celestial observation is used to locate and describe various surface characteristics. These specifications are defined in terms of latitude, time, longitude, and azimuth.

Observations of celestials require readings and measurements of horizontal and vertical angles; these readings are incorporated into plane and spherical triangles. The sextant and transit are two instruments commonly used to make these observations. Coupled with various reference tables, it is possible to locate the exact longitude and latitude of a point on the earth's surface.

A basic measure used in celestial observation is time. Time is categorized into three types of readings. The first is apparent solar time, which is similar to sundial readings. The second is sidereal time, which is based upon the hour angle of the March equinox at a specific point on the earth's surface. The third type of time is mean solar time, which is based upon the principle of the 24-hour day.

Time can be specified as either standard time or local mean time. Standard time is used for larger areas of land and is based upon a reference point at Greenwich, England known as Greenwich Mean Time. Local mean time pertains to the exact time on the meridian at which the observation point is located.

Transits can be used to make various celestial observations. The two basic observations made are sun or solar, and star. Sun observations are made to determine altitude and azimuth. Readings are made on the sun by its relationship to the cross hair lines in the transit. Star observations are easier to make through the transit because they appear as light points in the sky, allowing their placement on the cross hair lines to be more precise. Time can also be determined by use of transits. Field observations are conducted for time when the standard time is unknown, or if the local mean time is required. Again, time can be determined by solar or star observations.

Latitudes are used to compute time and azimuth, and can be obtained by any one of three celestial methods. These are by the sun's altitude at noon, by circumpolar stars, or from Polaris at any hour angle. Longitude can be obtained by scaling or by calculating the difference between time for a known longitude and an unknown longitude.

An azimuth of a line is a specification in relation to a reference meridian. The celestial observation techniques used for determining azimuth are observation on Polaris at elongation, by Polaris at any hour angle, by solar observation, solar attachment, and equal angles.

Surveying teams usually follow a specific sequence of making celestial observations and calculations. The specification that is generally noted first is time, followed by latitude, and then longitude.

KEY TERMS

Apparent Solar Time
Celestials
Center-tangent Method
Circumpolar Star
Disappearing-segment
  Method
Eastern Elongation
Ephemeris Tables
Equal Altitudes Method
Greenwich Mean Time
Hour Angle
Latitude

Local Apparent Noon
Local Civil Time
Local Mean Time
Longitude
Mean Solar Time
Mean Sun
Polaris
Quadrant-tangent Method
Radio Time
Reticule
Roelof's Solar Prism

Sextant
Sidereal Time
Solar Attachment
Solar Circle
Solar Observation
Standard Time
Time
Transit
Trigonometric Tables
True Solar Time
Western Elongation

REVIEW

1. Why are celestial observations and calculations important to map and topographic drafting?

2. Explain the difference between a plane triangle and a spherical triangle.

3. Briefly discuss why apparent solar time should not be used during field surveying.

4. Briefly explain the difference between sidereal time and mean solar time.

5. What is the difference between standard time and local time? What is the similarity in the calculations for each?

6. What is radio time, and what is its usefulness to the field surveyor?

7. Briefly explain how sun observations are used to determine altitude and azimuth specifications.

8. In your own opinion, are solar observations or star observations easier to observe through the transit? Explain.

9. Outline the various techniques that can be used to calculate time by transit observation.

10. Explain how latitude and longitude for a point can be plotted on an existing map.

11. What is a circumpolar star, and how can it be used to determine latitude?

12. Explain the difference between eastern and western elongations.

13. Briefly describe the difference between azimuth specifications by solar observation and by solar attachment.

14. What is the principle involved in the use of equal altitudes of the sun or a star to compute azimuth?

15. What are the three basic observations made in field celestial surveying? Specify their appropriate sequence for observation.

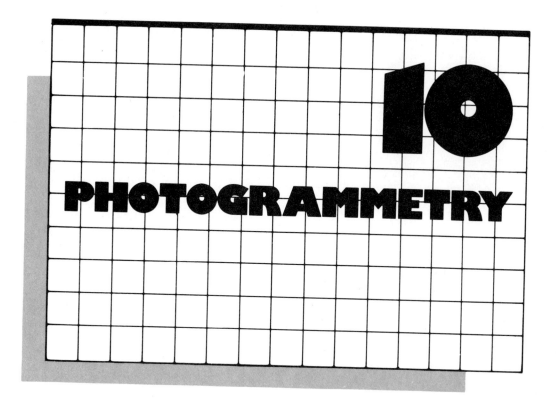

10 PHOTOGRAMMETRY

The principles of photogrammetry can be traced back to the works of Leonardo da Vinci in the late fifteenth century. Through his work in the fields of geometry, optics, mechanics, and geophysics, da Vinci graphically illustrated the concepts of aerodynamics and optical projection. It was not until the mid-1800s that the field of photogrammetry came into its own, particularly within the European community. The term *photogrammetry* was first coined in 1855 by the European geographer Kersten, and was in common use throughout Europe by 1900. The term, however, did not win wide acceptance in the United States until 1934. In that year, the American Society of Photogrammetry (ASP) was founded.

10.1 TYPES OF PHOTOGRAMMETRY

The ASP currently considers photogrammetry to be the art, science, and technology of surveying and measuring by photographic and other energy-emitting processes. The procedure is extensively used in topographic mapping, and can be applied from the ground as well as in the air. Aerial photogrammetry, though, is the procedure most commonly used in mapping. It has all but eliminated the need for extensive field surveys. Ground, or terrestrial, photogrammetry is used only as a supplement and complement to the aerial process, or in areas with unusual physical characteristics.

Like other professional disciplines, photogrammetry consists of a variety of specialty areas. For example, there are the two broad areas of *terrestrial* (ground) and *aerial* (air) *photogrammetry*. Other specialty areas of photogrammetry are defined as to the technology or energy-emitting process used. Some of the types of photogrammetry used in the mapping and topographic field are discussed here.

Radargrammetry is the use of radar as a measuring device to describe the physical characteristics of an area of the earth's surface. *X-ray photogrammetry* employs X-rays to collect surveying data, and the use of motion pictures for surveying is called cinephotogrammetry.

Hologrammetry is the use of holographs (images projected by the use of coherent light systems such as lasers) to measure surface characteristics. In *monoscopic photogrammetry*, single images or photographs are used for surveying purposes. Finally, there is space or satellite photogrammetry, in which spacecraft or satellites are utilized as a platform for taking surface measurements. This type is also referred to as extraterrestrial *photogrammetry*.

10.2 AERIAL PHOTOGRAPHY

The aerial photograph is perhaps the most familiar product of the photogrammetry field. Its application in the mapping and topographic data collection process combines the use of scientific and artistic procedures and techniques. Aerial photogrammetry, then, is the use of photographic images taken from an airborne base for surveying purposes.

The success and accuracy of an aerial photography mission is dependent upon a number of important factors, including the following:

- The use of correct photographic equipment, lenses, and supplies
- Appropriate selection and application of photographic materials, such as film, print and duplicating materials, reflection print materials, and plates
- The photographic team (pilot, photographer, and film processor)
- Weather conditions
- The position of the sun when the photographs are taken

Classifications of Aerial Photography

Aerial photography is a complex and dynamic profession, as is the photogrammetric field in general. Aerial photography requires a vast array of equipment, technologies, and procedures. Therefore, to classify the various types of aerial photographs used in the mapping and topographic field, the ASP has identified four criteria: the orientation of the camera axis, the lens system, the spectral range, and the mode of scanning. These criteria are discussed in this section.

Orientation of Camera Axis. This refers to the angle at which the camera is intentionally positioned at the time of the photograph. As shown in Figure 10-1, there are two orientations of axis used in aerial photography. These are the *vertical orientation axis* and the *oblique orientation axis.*

With a vertical orientation axis, the photograph is taken when the camera is intentionally positioned as nearly vertical — or as close to 90° to the earth's surface — as possible. Because of the

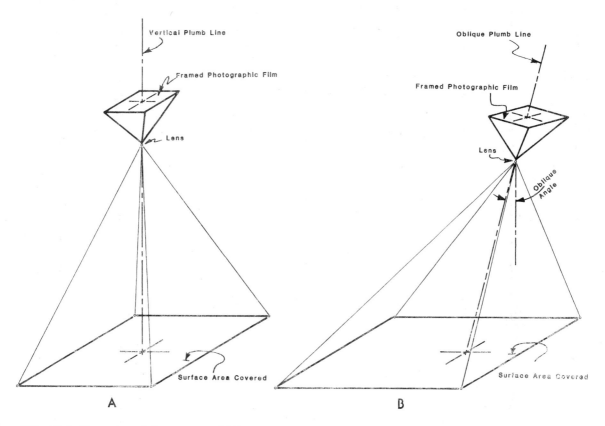

FIG. 10-1 Conceptual illustration of (A) vertical orientation axis and (B) oblique orientation axis

earth's curvature and aeronautical factors, it is impossible to obtain a true theoretical vertical picture (hence the qualifier "as nearly vertical as possible"). For practical mapping purposes, however, the "near vertical" is more than sufficient.

The vertical orientation axis is the most common type of aerial photograph used in mapping. As shown in Figure 10-2, this type of photograph can be easily converted into map drawings; it gives a sense of an existing map or chart. The three major advantages of using the vertical orientation axis are that measurements can be easily taken off the photograph and transferred, surface objects and landmarks can be easily identified, and the amount of hidden ground

(areas not observable in the photograph) will be minimal.

Oblique orientation axis photographs are images intentionally registered off vertical, and on the oblique. Technically, this means that the camera lens angle is aimed between true horizontal and vertical. These photographs can be further subclassified into *high oblique* and *low oblique*. High-oblique photographs pertain to the optical axis of the camera being at a high angle to the vertical, while low-oblique photographs have the optical axis of the camera at a low angle to the vertical.

Oblique photographs have a limited application for mapping, Figure 10-3. They are only of practical use in situations re-

FIG. 10-2 Aerial photograph using vertical orientation axis *(Courtesy of U.S. Army Corps of Engineers)*

quiring small-scale drawings. The two advantages of the oblique photographs are that they provide a stereoscopic or three-dimensional perception of the photographed area, and they can provide a larger area coverage than is provided by vertical photographs. There are, however, three significant disadvantages to using these photographs for map and topographic drawings. First, there is a loss of imagery and resolution. Second, the scale

constantly reduces as one observes images away from the camera. The third disadvantage is that highly skilled personnel and specialized equipment are needed to generate an accurate and useful product, resulting in higher costs.

Lens Systems. These are classified by the number and configuration of the camera lenses used for aerial photographs. Within this framework, all lens systems can be

FIG. 10-3 Aerial photograph using oblique orientation axis *(Courtesy of U. S. Army Corps of Engineers)*

divided into two broad classifications: the *single-lens systems,* and the *multiple-lens systems.*

The single-lens system is the most popular and frequently used system in aerial photography. It is applicable for either vertical or oblique photographs. The single-lens system usually employs a 153 millimeter (mm) focal length with a 228 × 228 mm format, while using only one lens per shot. By comparison, super wide-angle lens cameras are seldom used. If used, however, it is for photographic missions over low-relief terrain.

Multiple-lens systems consist of two or more lenses. These systems can be sub-classified by the type of mounting used. The first mounting is where the lenses are separately mounted in different cameras and the shutters are synchronized for simultaneous exposures. In the second mounting, the lenses are mounted inside the same camera body and share the same shutter for simultaneous exposures. In each case, the lenses are fixed at different optical axes so that the photographs can be used on a stereoscope. The accuracy and usefulness of photographs taken with a multiple-lens system are dependent upon the care and accuracy of the calibration and retention of the axis angles.

Another term used for the multiple-lens system is *multispectral.* Multispectral systems consist of two or more cameras, simultaneous exposures of an area of land, and the choice of different films and/or filter combinations.

Spectral Range. This criterion refers to the entire range of light on the spectrum. The vast majority of aerial photography is limited to the range of the spectrum that can be observed by the human eye. This is called the *optical range.* In some situations, aerial photographs are used to record images within the infrared range.

Spectral range is measured in terms of micrometers (μm). As shown in Figure 10-4, the optical range is from $0.4\ \mu$m to $0.8\ \mu$m, while the infrared range starts at $0.8\ \mu$m. With the use of photographic film capable of recording optical range images, it is also possible to record some infrared images at the lower end of the scale (*i.e.,* $0.8\ \mu$m and $0.9\ \mu$m). Photography further into the infrared range requires the use of special film and procedures. Because infrared is used to photograph temperature differences, it is also referred to as *thermal photography.* Figure 10-5 shows an example of infrared photography.

Mode of Scanning. This is the last criterion used in classifying aerial photographs.

There are three scanning modes used in aerial photography: the *single frame, panoramic,* and *continuous strip.* In this sense, *scanning* refers to how the camera lens functions during the photographic process.

Of the three scanning modes, the single frame is the most common and most desirable in the mapping and topographic field. In the single-frame camera, the entire frame (format) is exposed through a lens that is fixed in relation to the focal plane. These cameras are the easiest to use and the least expensive. They also provide the most accurate data to transfer onto maps and topographic drawings. There are, however, two disadvantages to single-frame photographs. First, there is a restricted field of view, and second, there is a decrease in resolution as one moves away from the center of the photograph.

The problems associated with single-frame photographs can be solved with the use of panoramic or continuous-strip cameras. Panoramic photography combines a wide image area with high resolution. This method maintains a high resolution at the center of the picture over the total angle scanned, which in some cases can be from horizon to horizon. Panoramic photography is accomplished by using narrow-angled, fast lens systems, by scanning the lens system through large angles across the flight path, and by using normal-width film and advancing it parallel to the scanning direction at ground speed.

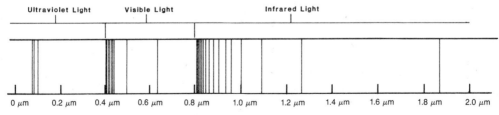

FIG. 10-4 Spectral ranges of light

FIG. 10-5 Infrared photograph *(Courtesy of U.S. Army Corps of Engineers)*

Continuous-strip photography eliminates the need for a conventional shutter system. The images are exposed on the film as the film passes continuously over a narrow slit in the focal plane of the lens at ground speed. Ideal for low-altitude reconnaissance photography, continuous-strip photographs provide a sharp picture scanned in one long strip.

Ground Control Points

The application of ground control points is a system used to ensure the accuracy of aerial photograph interpretation. *Ground control points* represent a series of fixed references which establish positions and elevations. Ground control points are also used for correlating various map features. Control is classified into four *orders,* which represent the degree of accuracy and precision. The first order signifies the highest degree of precision or quality.

The precision of the control system is based upon the use of the following seven types of controls.

1. *Basic control* is determined in the field, and is permanently marked, or monumented. It is based upon horizontal and vertical control of the third order or higher.
2. *Horizontal control* is control relative to positions referenced to geographic parallels and meridians; that is, references with horizontal positions only (*e. g.,* latitude, longitude, or plane coordinate axis).
3. *Vertical control* is usually made in reference to sea level and ground elevation.
4. *Astronomical control* is control established by astronomical observations. (See Chapter 9, Celestials.)
5. *Geodetic controls* are those controls that account for curvature of the earth's surface.
6. *Ground control* is frequently associated with basic control and geodetic control. These are references that have been established by ground or field surveys.
7. *Supplemental control* is made when additional, or subordinate, surveys are conducted to correlate the aerial photograph with geodetic control, thus ensuring positive identification of ground features and monuments.

Applications of Aerial Photography

To the casual observer, it may appear that aerial photography is only good for providing pictures of the earth's surface from an overhead perspective. Nothing could be further from the truth. There are a number of applications for aerial pho-

tography in the mapping and topographic field. These applications are useful to people in various professions, including engineers, geologists, geographers, planners, city administrators, agriculture specialists, lawyers, and economists. All applications, however, fall into four major categories. These are photointerpretation, stereocompilation, orthophotography, and analytical aerotriangulation.

Photointerpretation is a process used to analyze aerial photographs for the identification and measurement of surface objects and features. When used in an engineering study, this process usually emphasizes the relationships between surface objects and features and the project itself. Any person who is trained in photointerpretation can obtain a substantial amount of information. When applied to specific professions, this information can mean the difference between financial success and failure.

Stereocompilation is the process of extracting information from a stereo model. A stereo model is a three-dimensional image or model that is formed when the projecting rays of an overlapping pair of photographs intersect. The three-dimensional model, then, is based upon the use of aerial photographs, taken at slightly different angles, of the same area of land.

The purpose of a stereocompilation is to extract precisely located feature information from aerial photography. The stereo model created is characterized by vertical "stretching" or exaggeration that emphasizes the difference of contour features, and makes it easier to produce accurate map features. Hence, stereocompilation makes it possible to provide great accuracy of geometric and geodetic control data.

Orthophotography is the use of an aerial photograph as the final map product. It represents an immensely useful substitute for small-scale maps. This process is widely used by geographers and by agencies such as the United States Geological Survey.

The process of orthophotography eliminates the need for transferring all measurements and information from the aerial photograph to a drawing. The map produced by orthophotography is called a *photomap*. The process begins with a perspective photograph. The displacements of all images due to tilt and relief are removed; this is known as *differential rectification*. All information not shown on the photomap, such as scale, names, and elevations, are either drawn, scribed, or overprinted during reproduction.

Analytical aerotriangulation is a procedure required to supplement ground control points which are too far apart for photogrammetric compilation needs. Analytical aerotriangulation means that the coordinates of photo control points are produced by mathematical procedures rather than by analogue methods. This is a highly technical and analytic procedure used to establish precise photogrammetric data.

10.3 SPACE PHOTOGRAMMETRY

Satellites were first launched and orbited in the late 1950s. As a result, an entirely new technology, space technology, became available to many established engineering and scientific disciplines. One professional area that was significantly affected was photogrammetry.

The advent of the space age caused many photogrammetrists to become intrigued by and interested in the potentials of recording surveying data from an orbiting base. Unlike aerial photography and photogrammetry, *space photogrammetry* is the science of using sensors in a spacecraft (such as a satellite, manned orbiter, or shuttle) to secure mapping and plotting coordinates and measurements. Space photogrammetry is used not only for mapping the earth's surface but also for planetary mapping. Thus, the equipment used in this field is suited for collecting mapping data for the earth as well as for other planets and their natural satellites or moons.

Space Photogrammetry Systems

Space photogrammetry systems are classified according to their application. All space sensing systems can be categorized into one of two applications: *remote sensing* or *satellite photogrammetry*. Remote sensing is based upon the concept that the characteristics and nature of surface objects and terrain are more important in data collection than is the geometry of these features. Satellite photogrammetry, on the other hand, is just the opposite; the geometry of surface objects and terrain is more important than their nature and characteristics. Satellite photogrammetry can be further divided into the two subcategories of *planimetric photogrammetry* and *topographic photogrammetry*.

In planimetric or two-dimensional space photogrammetry, the height measurements of surface objects are not recorded for analysis. The planimetric system is used because the high altitude of a spacecraft makes it difficult to accurately record and interpret surface height measurements. Examples of planimetric satellite systems are Landsat and weather satellites.

In topographic or three-dimensional photogrammetry, surface height measurements are recorded for analysis and interpretation. The topographic system uses a stereoscopic imagery system and is specifically designed for photogrammetric geodesy. This technique provides information on a three-dimensional coordinate axis, and is particularly suited for the generation of topographic products. Figure 10-6 presents a conceptual configuration of how the three-dimensional stereo model is obtained.

Elements of Space Photogrammetry

It is assumed in space photogrammetry that all images and data recorded will be directly along the spacecraft's orbiting path. Therefore, the first critical element to determine is the craft's position in the orbit at the exact time that data are recorded. Because of the high altitudes used, it is impossible to use only the recorded images for this purpose, for the area they cover is too large. Therefore, earth-based tracking stations are used to determine the exact orbit location. The stations are located in key positions around the world so that the spacecraft can be tracked 24 hours per day.

These tracking stations also collect other data pertinent to the photogrammetric mission, such as spacecraft direction, distance, and velocity. Elements necessary to determine for successful space photogrammetry are the size and shape of the orbit, the orientation of the spacecraft, and the exact instant that the spacecraft reaches perigee (nearest point to the planet's surface).

10.4
MOSAICS

A single photograph recorded from an airplane or spacecraft has image and measurement distortions as the viewer moves away from the center of the photograph. Because of these distortions, it is necessary to use more than one photograph. On aerial and space photogrammetric missions, a series of photographs is taken. The photographs are then combined into one large photograph called a *mosaic*.

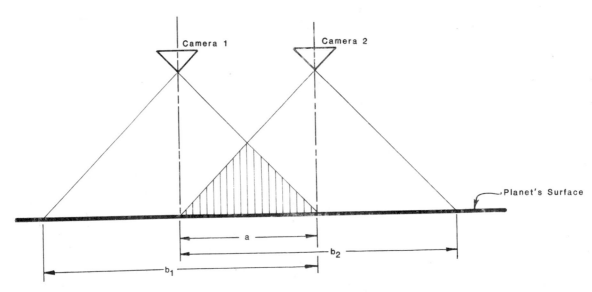

FIG. 10-6 Conceptual illustration for generating stereo model

A mosaic is the fitting together of individual photographs to provide a composite picture of an area that is recorded on an aerial or space photographic mission. A mosaic that is properly assembled should appear as a single photograph with minimal pictorial distortion. The production of a high-quality mosaic requires care, a systematic procedure, and some artistic talent. The mosaic process includes five essential steps. These steps are scaling, photography, assembling, blending, and annotating.

Scaling

In the mosaic process it is first necessary to determine and specify an appropriate scale. The scale should be identified prior to taking the photographs. Factors affecting scale selection are use, size and shape, and the financial resources. As shown in Table 10-1, the larger the scale selected, the more applicable will be the mosaic for engineering projects.

Photography

Once the scale has been determined, it is possible to plan and execute the photographic mission. An important element of the entire photographic process is the use of control points. *Control points* are predetermined locations, or points, over the area to be photographed. These points are used as fixed references when the mosaic is assembled. Figure 10-7 shows two examples of control-point configurations used for a photographic mapping mission.

Another important element of photography is the amount of *overlap*. This refers to how much each sequential image overlaps the others, Figure 10-8. The amount of overlap is directly related to the accuracy and amount of distortion in the mosaic — as the amount of overlap increases, accuracy increases and distortion decreases. For most photographic missions, overlap ranges between 30% to 60% (30% between adjacent flight lines and 60% along flight lines).

Table 10-1
Mosaic Scales

Scale Size	Ratios	Examples of Use
Small	1 : 20,000 and smaller	Geology, forestry, flood control, military, and reclamation
Medium	1 : 10,000 to 1 : 20,000	City planning; preliminary work for the location of highways, waterways, railroads, and powerlines or utilities
Large	1 : 10,000 and larger	Detailed engineering work for projects such as mining, highway, railroad, residential, and commercial development construction

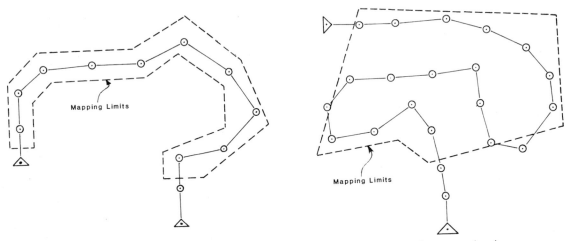

FIG. 10-7 Control point configuration for two types of maps. NOTE: The photographed area exceeds mapping limits in all cases.

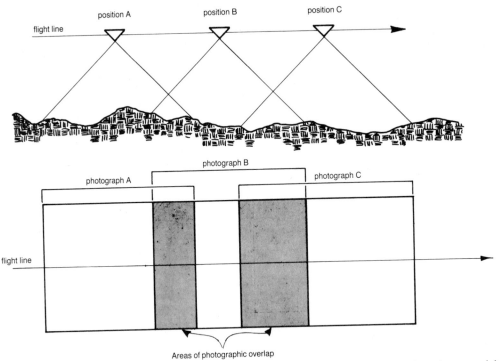

FIG. 10-8 Photographic overlapping. NOTE: The amount of photographic overlap that would appear from images recorded through the same camera art positions a, b, and c is shown.

Assembling

Assembling is the "putting together" of the mosaic. To ensure a high-quality prod-

uct, mosaics should be assembled on a layout board or assembling table which has a hard and smooth surface. A well-constructed board or table should be made of masonite, plywood, chipboard, or aluminum, and should be large enough to handle most of the projects encountered. (Mosaics used for engineering purposes are prepared to the same sizes as engineering drawings.)

Once the layout board is ready, the center photograph is located and mounted in the center of the board, followed by the alignment of photographs of contiguous areas. (Figure 10-9 shows a typical sequencing of photographic alignment used in mosaics.) A fine needle is used to prick a hole through the identified control points to ensure that each photograph is properly aligned. The overlapping photographs are then positioned by aligning the control-point holes over one another.

As each print is positioned into place it is automatically mounted and stabilized. This is accomplished by applying an adhesive to the board itself or to the back of each photograph. Gum arabic, glue, rubber cement, and paste are some adhesives commonly used. Once all prints are mounted, excess adhesive is wiped off and the mosaic is trimmed.

Blending

Blending is a finishing process used to make sure that each photograph blends together with all other photographs. In this process, photographic dyes are manually applied to the photographs. To avoid excessive touching up, great care is given to the negative and print processing procedure. However, even under the best of conditions and highest-quality equipment and materials, some blending is usually required.

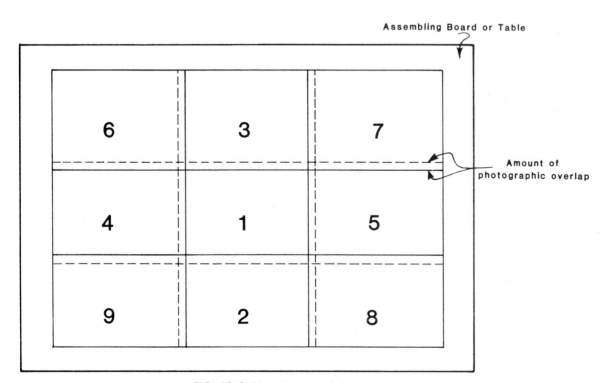

FIG. 10-9 Mosaic assembling sequence

Annotating

Mosaics often require more information than that presented on the photographs. The additional information is descriptive and annotated. Examples of annotated information added to mosaics are titles, scale, grid designations and references, cultural features, political boundaries, and property lines. Annotation is accomplished by mechanical, hand, and/or transfer overlay lettering. Once the annotated information is added, the mosaic is ready for reproduction.

10.5 PHOTOGRAMMETRY IN MAP DRAWING

The development and refinement of photogrammetric processes have almost eliminated the need for field or ground surveying to obtain mapping source data. Photogrammetry techniques that are applicable to map drawing range from the generation of simple, single-image aerial photographs to multisensory systems that record a vast array of topographic and surface data.

Producing an accurate map or topographic drawing requires the use of precise information and map drawing procedures. Utilizing photogrammetric data in map drawing requires three procedures: stereocompilation, drafting or scribing, and editing. (These procedures are discussed here only in relation to the use of photogrammetric data in map drawing.)

The Use of Stereocompilation

Stereocompilation is one of the most common methods of extracting topographic data from photographs. Put simply, stereocompilation is the generation of maps and topographic drawings, on a three-dimensional orientation, from aerial photographs. This process is accomplished with the use of stereo-plotting equipment, Figure 10-10. Stereocompilation is divided into four steps, which are aerotriangulation, model orientation, planimetry, and contours.

Aerotriangulation. This step includes all procedures required to ensure accurate and applicable source data, including the establishment of scales, azimuth, vertical data, and coordinate systems. An aerotriangulation procedure can range from simple references to control points to complex mathematical modeling through analytical aerotriangulation.

Model Orientation. This refers to orientation of aerial images in the stereo plotter so that stereoviewing is possible when two overlapping images are projected. The model itself consists of the three-dimensional imagery of the photographed area. The model is viewed through the stereo plotter, which projects the three-dimensional image by the intersecting of rays from two or more overlapping photographs. Figure 10-11 is a conceptual presentation of how a three-dimensional object is projected on a stereo plotter.

Planimetry. This is process of determining area measurements on the map model. Planimetric measurements can be made easily for open and undeveloped areas. Planimetry is much more time consuming and difficult, however, for developed areas (such as cities with a large number of streets, utilities, and buildings) or land that has a variety of terrain (such as mountains, lakes, and rivers).

In the context of stereocompilation, planimetry refers to map features other than

FIG. 10-10 Stereo-plotting Equipment. This system produces three-dimensional imagery required for examining topographic features and structures. *(Courtesy of Bausch & Lomb Scientific Optical Products Division)*

surface forms or elevation. Planimetric features such as streams, roads, and open bodies of water are delineated from the stereo model as a separate step. Contours are also identified in a separate step.

Contours. These are lines drawn to illustrate the connection of established points for a given elevation level. Depending upon the scale and use of the drawing, contour lines may be drawn for elevations varying from every 5 feet to every 20 feet. Maps used for engineering projects require more detailed contours. Therefore, there is a need for larger-scale drawings and smaller vertical intervals between contour lines.

the map is to be multicolored and reproduced, a process known as *separation drafting* is used. This technique employs a separate drawing per color. Hence, a four-color map requires four separate drawings. For single-color maps, only one drawing is prepared; this technique is the fastest and least expensive.

Generally, a separate plate is produced for each type of feature. For example, single-line streams and outlines of open-water bodies would be scribed on one plate. Open-water areas would be delineated on another plate and screened so that only one blue ink is required. This procedure is used for published maps.

Drafting or Scribing

The finished map is either drawn or scribed from the photogrammetric data. If

Editing

Editing is the checking of the map at various stages of map drawing and data trans-

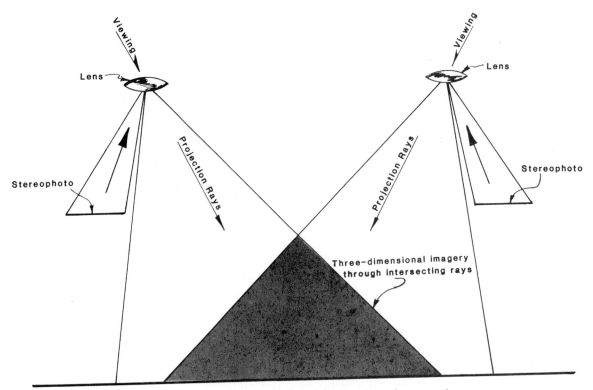

FIG. 10-11 Conceptual illustration for three-dimensional viewing of stereo plotter

fer. This is a highly important procedure in the map production process.

Editing ensures map accuracy, completion, and data interpretation. It also provides for a quality-control check to make sure that the drawing is legible and can be reproduced. Editing during the map drawing phase is referred to as either a *scribing edit* or *drawing edit.*

10.6 SUMMARY

Photogrammetry became an accepted and widely used term in the United States in 1934, when the American Society of Photogrammetry was founded. Photogrammetry is a process used for surveying and measuring the surface features of the earth through various energy-emitting processes. These processes can be used to describe various specialty areas within the profession. Some of the types used in mapping and topographic work are radargrammetry, X-ray photogrammetry, cinephotogrammetry, hologrammetry, monoscopic photogrammetry, and space or satellite photogrammetry.

The most common type of a photogrammetry product is the aerial photograph. Aerial photographs are classified according to four criteria: orientation of the camera axis, lens system, spectral range, and mode of scanning. The aerial photograph is useful in many mapping and topographic functions. Some of the major applications of aerial photographs are photointerpretation, stereocompilation, orthophotography, and analytical aerotriangulation.

The availability of space technology has resulted in the development of two major space photogrammetry systems—remote sensing and satellite photogrammetry. Satellite photogrammetry can be subdivided into planimetric and topographic measuring procedures which give two- and three-dimensional data.

Mosaic preparation is a specialty area in which aerial photographs are fitted together to provide a composite picture of an area of land. Mosaic preparation includes scaling, photography, assembling, blending, and annotating.

Photogrammetry has greatly reduced the need for ground surveys in the mapping process. Therefore, it is frequently used as the method for obtaining mapping source data. The use of photogrammetric source data in map drawing requires three procedures—stereocompilation, drafting or scribing, and editing.

KEY TERMS

Aerial Photogrammetry
Aerial Photography
Aerotriangulation
Analytical Aerotriangulation
Annotating
Assembling
Astronomical Control
Basic Control
Blending
Camera (Lens) Axis
Cinephotogrammetry
Continuous Strip
Contour
Control Point
Drafting
Editing
Extraterrestrial
 Photogrammetry
Geodetic Control
Ground Control

Hologrammetry
Horizontal Control
Lens System
Mode of Scanning
Model
Model Orientation
Monoscopic Photogrammetry
Mosaic
Multiple-lens System
Oblique Orientation Axis
Optical Range
Orthophotography
Panoramic
Photogrammetry
Photograph
Photointerpretation
Photomap
Planimetric
 Photogrammetry
Planimetry

Radargrammetry
Remote Sensing
Satellite Photogrammetry
Scaling
Scribing
Separation Drafting
Single-frame Image
Single-lens System
Space Photogrammetry
Spectral Range
Stereocompilation
Stereo Plotter (Stereoscope)
Supplemental Control
Terrestrial Photogrammetry
Topographic Photogrammetry
Vertical Control
Vertical Orientation Axis
X-ray Photogrammetry

1. In your own words, explain photogrammetry and its uses in mapping and topographic drawing.
2. Identify and describe the two broad areas of photogrammetry.
3. What is aerial photography?
4. Discuss the importance of the following factors in aerial photography:
   a. Orientation of the camera axis
   b. Lens system
   c. Spectral range
   d. Mode of scanning
5. Describe the difference between high-oblique and low-oblique photographs.
6. Briefly explain how single-lens and multiple-lens systems function in aerial photography.
7. Define the term *optical range*, and explain its importance to aerial photography.
8. Describe panoramic photography.
9. Describe these processes:
   a. Photointerpretation
   b. Stereocompilation
   c. Orthophotography
   d. Analytical aerotriangulation
10. What is a photomap?
11. How does space photogrammetry differ from other types of photogrammetry?
12. Describe the difference between planimetric photogrammetry and topographic photogrammetry.
13. Explain how orbiting satellites are kept in constant communication with the earth.
14. What is a mosaic?
15. What are control points, and what function do they serve in producing mosaics?
16. Explain how photographic overlap can increase the accuracy of a mosaic.
17. Why are mosaics annotated?
18. Briefly describe the process of assembling a mosaic.
19. What is a model, and how does it function during stereo-compilation?
20. Briefly explain separation drafting and its function in mapping and topo drawing preparation.

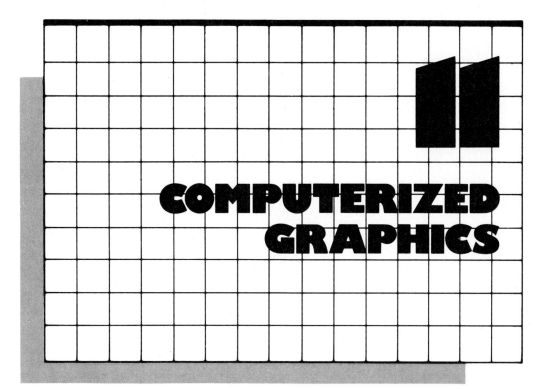

COMPUTERIZED GRAPHICS

Computers and computer systems have become commonplace in helping engineers, geographers, cartographers, and geophysicists produce topographic drawings. Until a few years ago, the use of computerized graphics was limited to large firms and government mapping agencies that could afford to purchase or rent computers. In addition, these firms limited their use of computers to storing large quantities of data and calculating complicated mathematical formulae.

Advancements in microcircuitry and computers have increased the availability and use of computers. No longer are these systems affordable only to large businesses with extensive budgets. Computerized graphics are now within reach of small businesses and the individual professional.

Not only have computers and computerized graphics become affordable, but their use is now more of a necessity than a luxury.

11.1
THE COMPUTER GRAPHICS PROCESS

Computer graphics is the application of computer systems to the production of graphic displays. *Computer systems,* in this context, refer to the *hardware* (physical equipment) and *software* (programs) used

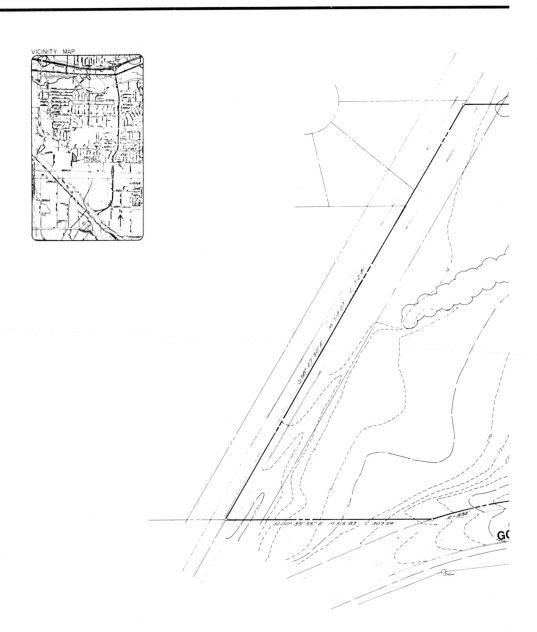

CONCEPT PLAN
GOODLETT ROAD DUPLEX SU

MEMPHIS, TENNESSEE   PREPARED FOR: ATLANTIC SOUTHERN CORP

FIG. 11-1 Contour topographic drawing prepared on plotter. All line work, excluding lettering, was

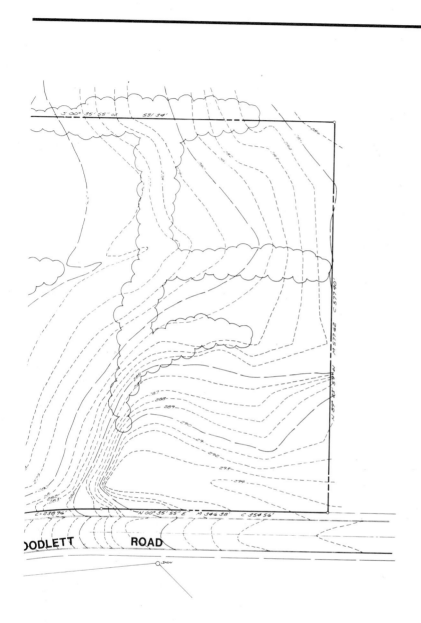

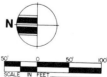

BDIVISION

ORATION

printed on a plotter. (*Courtesy of Gregory-Grace and Associates, Inc.*)

in generating a graphic product. Depending upon the hardware used, graphic displays can be presented in single color or multicolor, as two- or three-dimensional presentations, and as an image on a viewing screen or a hard copy on a sheet of paper or film.

The line drawing is the most common and least expensive form of computer graphics. Most drawings used in the preparation of topographic drawings are line drawings. They are therefore more closely associated with computerized graphics. Figure 11-1 shows a topographic computerized line drawing.

The pictorial drawing is another, and more sophisticated, form of computer graphics. Pictorial drawings are used to present life-like illustrations of surface features. Their advantage over line drawings is understandability. They are easier to interpret, and can be visualized by the untrained person.

Perhaps the greatest advantage of using computers is their ability to generate data and drawings of contours resulting from changes in the land surface. Based upon mathematical modeling, it is possible to "see" what will happen to land if the contour is changed (as a result of excavation or filling). This is a critical function in engineering, since many projects are labeled as either acceptable or not acceptable based upon their effect on the topography, and how that topography affects the physical features and mechanics of the land (*i.e.*, drainage, erosion, traffic lines of sight). Hence, different scenarios can be run through a computer to determine the best contour for a given piece of land with a specific usage.

Utilization of computers and computerized graphics requires that the operator be familiar with the processes or system used to generate a product. There are four basic processes that must be sequentially followed to generate a drawing and/or calculation. These processes are modeling, translation of data, data input, and data output.

Modeling

A model is an estimate or approximation of the real world. Within the realm of computer science, *models* are mathematical formulae used to explain or predict a phenomenon. For example, an existing contour can be explained or estimated mathematically based upon its elevation, length, and area. Likewise, an ideal or model contour can be developed that will provide the best surface layout for a piece of land, to allow adequate drainage, reduce erosion, and promote vegetative growth.

With the calculations of the existing contour and the ideal contour, it is possible to calculate the amount of excavation or fill needed. As shown in Figure 11-2, the computer printout identifies areas of land to be removed and those to be filled.

Translation of Data

The next process, translating topographic data into acceptable computer form, is critical. If data cannot be "read" by the computer, the data cannot be interpreted nor analyzed. Instructions to the computer for processing the data are conveyed in a format known as *computer languages*. Examples of these languages are BASIC, COBOL, and Fortran. All computer languages are based upon a logical and systematic order of giving instructions to the computer, and sequencing of data. The instructions or directions tell the computer what is to be done with the data, and the

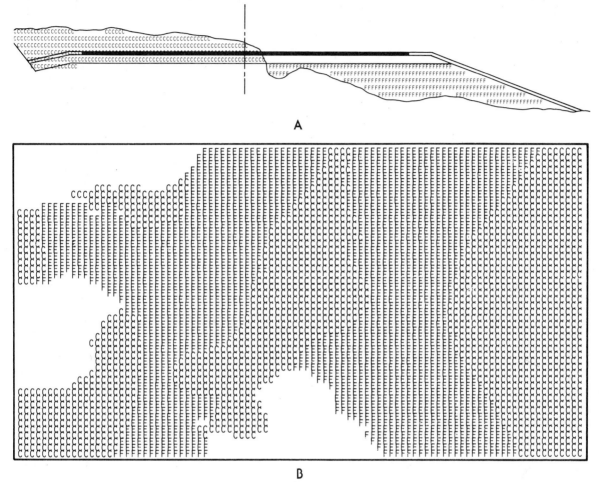

FIG. 11-2 Computer printout showing cut and fill areas. (A) A profile printout, (B) a plan view printout for a different area.

type of product desired (drawings, tables, and so on).

Figure 11-3 is an example of computer language usage to generate a graphic display. Notice that the instructions are short and in a sequential order.

Data Input

The process of placing the data into the computer is referred to as *data input.* Input consists of the data itself, along with the instructions of what is to be done with the

data. These directions are called *programs.* Programs can be written by the individual user, and can be tailored for a particular use. Other programs can be purchased prewritten, and are known as *canned programs.* The programs, then, make up the software for the computer system.

Data Output

Computer systems are made up of a variety of hardware (card readers, commu-

```
999 SELECT LIST 215
1000 $PSTAT= "SURVEY"
1140 COM D$6,0,Q,Y(14),X(14),Z0: DIM O$64,M$64,T$1,NO$(28)32,A$(12)3: SELECT DIS
K 320,#1B10,#2B10,#3D10,#4320,#5320,#6320,#7D20,D: Z0=3
1190 DEFFN'15: IF D$<>" "THEN 1510
1191 GOSUB 6000: GOSUB '253(2,22,"ENTER TODAY'S DATE")
1240 GOSUB '255(3,24,14): IF O$=" "THEN 1240: IF STR(O$,1,1)="L"THEN 3000: IF ST
R(O$,1,1)="P"THEN 5000: IF NUM(O$)=0THEN 1240
1241 CONVERT STR(O$,1,NUM(O$))TO M: IF M<=0THEN 1240: IF M>12THEN 1240
1242 K=POS(O$="/"): IF NUM(STR(O$,K+1))=0THEN 1240: CONVERT STR(O$,K+1,NUM(STR(O
$,K+1)))TO D: IF D<=0THEN 1240: IF D>31THEN 1240: K=POS(STR(O$,K+1)="/")+K: IF K
=0THEN 1240: IF NUM(STR(O$,K+1))=0THEN 1240: CONVERT STR(O$,K+1,NUM(STR(O$,K+1))
)TO Y: IF Y<=0THEN 1240: IF Y<100THEN 1470
1450 Y=100*(Y/100-INT(Y/100))
1470 CONVERT MTO STR(D$,1,2),(##): CONVERT DTO STR(D$,3,2),(##): CONVERT YTO STR
(D$,5,2),(##)
1510 PRINT HEX(03): GOSUB '253(8,12,"LOADING SURVEYING MENU"): LOAD DC F "D.M.S.
"
1600 DEFFN'254(R,C)
1610 SELECT PRINT 205: PRINT HEX(01):: IF R=1THEN 1690: INIT(00)O$: INIT(0A)STR(
O$,1,R-1): PRINT HEX(0E):O$:
1690 IF C=1THEN 1740: INIT(00)O$: INIT(09)STR(O$,1,C-1): PRINT O$:
1740 SELECT PRINT 005: RETURN
1860 DEFFN'255(R,C,L): GOSUB '254(R,C): INIT(00)O$: INIT(2D)STR(O$,1,L): PRINT O
$: GOSUB '254(R,C): O$=" ": I=1: C1=C
1990 KEYIN T$,2030,1990: GOTO 1990
2030 IF T$=HEX(08)THEN 2140: IF T$=HEX(0D)THEN 2120: PRINT T$:: STR(O$,I,1)=T$:
I=I+1: IF I<=LTHEN 1990
2120 RETURN
2140 IF I=1THEN 1990
2160 PRINT HEX(082D08):: I=I-1: STR(O$,I,1)=" ": GOTO 1990
2270 DEFFN'253(R,C,M$): GOSUB '254(R,C): INIT(00)O$: INIT(20)STR(O$,1,64-C-1): P
RINT O$: GOSUB '254(R,C): PRINT M$:: GOTO 1610
3000 SELECT PRINT 005(64): INIT(" ")NO$(): I=0
3010  DATA LOAD DC OPEN T#1,"CATALOG": DATA LOAD DC #1,S1$
3020  PRINT HEX(0A):" DISK # ";S1$:" ": I=I+1: I1=1: IF I>28THEN 3050: DATA LOAD
  DC #1,NO$(I): I=0
3025 I=I+1: I1=1: IF I>28THEN 3050: DATA LOAD DC #1,NO$(I)
3030  IF ENDTHEN 3050: I1=0: GOTO 3025
3050  FOR I=1TO 28: PRINT STR(NO$(I),1,31):: NEXT I: PRINT
3080  KEYIN T$,3090,3090: GOTO 3080
3090 IF I1<=0THEN 3020: GOTO 1191
4090 RETURN
5000 PRINT AT(6,0):HEX(0E):"SELECT PRINTER BY NUMBER"
5001 %  ####################################
5002 PRINTUSING 5001,"No. ----------------------------------------"
5030 PRINTUSING 5001,"1    LOCAL WRITER   ( 204 )"
5040 PRINTUSING 5001,"2    SYSTEM LINE PRINTER    ( 215 )"
5050 PRINTUSING 5001,"2    SYSTEM DAISYWHEEL PRINTER ( 216 )"
5055 PRINTUSING 5001,"3    SYSTEM LINE PRINTER    ( 215 )"
5060 PRINTUSING 5001,"4    SYSTEM TYPE-WRITER   ( 213 ) ": PRINT HEX(0A0E):
```

FIG. 11-3 A short program for computer graphics. (*Courtesy of Gregory-Grace and Associates, Inc.*)

```
5070 Z0$="2": LINPUT "ENTER PRINTER NUMBER (1 - 4) ",Z0$: IF NUM(STR(Z0$,1,1))=1
THEN 1191: CONVERT STR(Z0$,1,1)TO Z0: GOTO 5070
6000 PRINT HEX(03): PRINT HEX(03);"********************** SURVEYING *********
********************": PRINT "**";TAB(62);"**": PRINT "**";TAB(62);"**"
6001 PRINT "** P -SELECT PRINTER ********************* L -LIST DATA DISK **": R
ETURN
10 REM COGO: DIM B(28),A$36,K$64: M=28: F1,F2,B1,B2,D1,D2,N1,N2,E1,E2,P1,P2=#PI:
  U=1/#PI: GOTO 2230
30      Q1=1+INT((Y7-1)/14)
31      IF Q1=Q THEN 43
32      Q=Q1-Q
33      IF Y8=0 THEN 37
34      DBACKSPACE #1,1S
35      DATA SAVE DC #1,Y(),X()
36      Y8=0
37      IF Q>0 THEN 40
38      DBACKSPACE #1,(ABS(Q)+1)S
39      GOTO 42
40      IF Q=1 THEN 42
41      DSKIP #1,(Q-1)S
42      DATA LOAD DC #1,Y(),X()
43      Q=Q1
44      Q1=Y7-14*(Q-1)
45      RETURN
330 PRINT TAB(80):TAB(80):TAB(80):
340 PRINT HEX(0C0C0C):: RETURN
350 SELECT PRINT 005: RETURN
360 PRINT HEX(0C): Z=Z-1
370 DEFFN'32(Z): IF Z>0THEN 360: FOR H=WTO 0STEP -1: GOSUB 1020: GOSUB '33(F): P
RINT TAB(6):: GOSUB '36(B,0): GOSUB '38(20,D): GOSUB '38(30,N): GOSUB '38(42,E):
 PRINT TAB(57):: GOSUB '33(P): PRINT TAB(80): NEXT H: RETURN
400 %###
420 RETURN
450 %#
460 % 0#
470 DEFFN'36(X,Z): R=X: IF X=#PITHEN 420: IF X=UTHEN 570: IF Z=0THEN 480: B$=" "
: IF X>=0THEN 490: B$="-": GOTO 490
480 B$="S": X=ARCSIN(ABS(SIN(X-360*INT(X/360)))): IF COS(R)<0THEN 490: B$="N"
490 GOSUB 520: X=ABS(X+SGN(X)/72E3): FOR V=1TO 3: IF X<10THEN 500: PRINTUSING 40
0,X:: GOTO 510
500 PRINTUSING 460,X:
510 X=(X-INT(X))*60: NEXT V: PRINT ".":: PRINTUSING 450,X/6:: IF Z<>0THEN 520: B
$=" W": IF SIN(R)<0THEN 520: B$=" E"
520 PRINT B$:: B$=" ": RETURN
530 %-#######.###
540 %-#####.###
550 DEFFN'38(Z,X): PRINT TAB(Z):: IF X=#PITHEN 420: IF Z=20THEN 560: IF X=UTHEN
570: PRINTUSING 530,X+5E-4*SGN(X):: RETURN
560 IF X=UTHEN 580: PRINTUSING 540,X+5E-4*SGN(X):: RETURN
570 PRINT "   ";
580 PRINT "   UNKNOWN":: RETURN
```

FIG. 11-3 (Continued)

```
590 DEFFN'41(Y): GOSUB 1980: PRINT TAB(Y);HEX(0C);: PRINT A$;: Z=#PI: INPUT Z: I
F INT((Y-4)/50)*INT((ABS(Z)-1)/0)<>0THEN 590
600 A$=" ": RETURN
605  REM GET CO-ORDINATE
610    DEFFN'44(X)
615    Y7,X=INT(ABS(X))
620    GOSUB 30
696    Y=Y(Q1)
697    X=X(Q1)
698    RETURN
700 DEFFN'45(X,Z): GOSUB '44(X): IF Y=#PITHEN 420: B=X: D=Y: GOSUB '44(Z): IF Y=
#PITHEN 420: X=X-B: Y=Y-D: D,B=SQR(X*X+Y*Y): IF D=0THEN 702: B=90*SGN(X): IF Y=0
THEN 702: B=ARCTAN(X/Y): IF Y>0THEN 702: B=B+180
702 IF B>=0THEN 703: B=B+360: GOTO 702
703 IF B<360THEN 704: B=B-360: GOTO 703
704 IF B<>0THEN 705: B=1E-50
705 RETURN
710 D=#PI: GOSUB '41(6): B=U: IF Z=#PITHEN 420: IF ABS(Z)>=1E6THEN 720: V=0: IF
ABS(Z)>=1E3THEN 730: IF INT((Z-1)/M)<>0THEN 710: B=INT(Z): Z=900*(Z-B): B=B(B):
IF D=#PITHEN 710: B=B+Z: RETURN
720 IF ABS(Z)<5E6THEN 750: R=Z/1E7: Z=(ABS(R)-INT(ABS(R)))*SGN(Z)*1E7: R=INT(ABS
(R)): B=0: GOSUB 770: IF R<100THEN 740: Z=R: V=X
730 R=Z-INT(Z): X=INT(INT(Z)/1E3): Y=INT(Z-X*1E3): IF INT((Y-1)/0)+INT((X-1)/0)<
>0THEN 710: GOSUB '45(X,Y): D=#PI: B=V+B+R*900: IF Y=#PITHEN 710: GOTO 760
740 IF R>MTHEN 710: IF B(R)=#PITHEN 710: B=B+B(R): GOTO 760
750 IF Z<0THEN 710: Y=INT(Z/1E6): X,Z=SGN(Y/2-INT(Y/2)-.3)*(Z-Y*1E6): B=INT(Y/2)
*180: GOSUB 770
760 A$="BRG#": GOSUB '41(22): IF Z=#PITHEN 420: IF INT((Z-1)/M)<>0THEN 760: B(AB
S(Z))=B: RETURN
770 X=Z: FOR Y=1TO 2: X=X/100: X=((ABS(X)-INT(ABS(X)))/.6+INT(ABS(X)))*SGN(X): N
EXT Y: B=B+X: RETURN
780 Y=31: GOSUB 1040: N=Z: Y=43: GOSUB 1040: E=Z: Z=P: Y=54: GOSUB 1050: P=Z
785 REM SAVE CO-ORDINATE
790    DEFFN'43(Z)
800    Y7,Z=INT(ABS(Z))
805    GOSUB 30
840    IF Z5<>0 THEN 900
850    Y(Q1)=N
860    X(Q1)=E
862    Y8=1
863    RETURN
900 IF Y(Q1)=#PITHEN 850: IF ABS(Y(Q1)-N)+ABS(X(Q1)-E)<.001THEN 850: KEYIN T$,70
0,900: PRINT TAB(80);HEX(0C);: PRINT "POINT #";: GOSUB '33(Z): INPUT " USED.. NE
W PT #...",Z: PRINT HEX(0C);TAB(80);HEX(0C);: IF INT(ABS(Z))=Y7THEN 850: P=Z: GO
TO 800
990 GOSUB 1000
1000 X=F2: F2=F1: F1=F: F=X: X=B2: B2=B1: B1=B: B=X: X=D2: D2=D1: D1=D: D=X: X=N
2: N2=N1: N1=N: N=X: X=E2: E2=E1: E1=E: E=X: X=P2: P2=P1: P1=P: P=X: RETURN
1010 F,B,D,N,E,P=#PI: RETURN
1020 W,H=H*W1: IF W=0THEN 350: SELECT PRINT 215: RETURN
1030 Y=54
```

FIG. 11-3 *(Continued)*

```
1040 A$=" ": GOSUB '41(Y)
1050 IF Z=#PITHEN 1040: RETURN
1060 X=N1: N1=N: N=X: X=E1: E1=E: E=X: X=F2: F2=P1: P1=X: B2=B1+180: D1=D2: RETU
RN
1070 N1=N: E1=E: F=P: P=P+1
1080 N=N1+D*COS(B): E=E1+D*SIN(B): RETURN
1090 N2=N+D2*COS(B2): E2=E+D2*SIN(B2): RETURN
1100 DEFFN'31: W=1
1110 DEFFN'15: PRINT HEX(03): L=0: FOR H=WTO 0STEP -1: GOSUB 1020: PRINT HEX(0D)
:: A$="ON": IF Z5<>0THEN 1120: A$="OFF"
1120 PRINT TAB(20);"** POINT PROTECT ";A$:" **": A$=" ": PRINT "FROM       BEARI
NG     DISTANCE      N COORD      E COORD      PT#": NEXT H: GOSUB 1000: GOSUB 1970:
GOSUB 1000: GOSUB 1970:
1150 GOSUB 1000: GOSUB 1010
1160 GOSUB 600: F=SGN(L)*(P1-#PI)+#PI: PRINT : IF L<>0THEN 1210
1161 PRINT HEX(0C);TAB(80);HEX(0C):: K$=" ": Z=#PI
1162 INPUT K$: IF K$=" "THEN 1168: IF NUM(K$)<>0THEN 1165: SELECT PRINT 215(80):
 PRINT K$: SELECT PRINT 005(64): GOTO 1161
1165 CONVERT STR(K$,1,NUM(K$))TO Z: IF Z>0THEN 1161
```

FIG. 11-3 *(Concluded)*

nication line switches, tape drives, printers, and plotters). Some hardware (viewing screens, printers, and plotters) will determine what form the display will take. Depending upon the type of display desired, as specified in the program, the topographic drawing will be either temporary (soft) or permanent (hard).

11.2 COMPUTERIZED DIGITAL GRAPHICS

Digital graphics pertain to those systems that give input in digital form. The physical materials used to provide digital input are punch cards, punched tape, and magnetic tapes. The information given can be either straight binary or binary-coded decimal.

Computerized digital graphic systems require little or no human intervention during the drawing stage of the program. These systems are quite flexible in terms of the type of stylus (drawing element) and

medium (drawing surface) that can be used. Topos can be prepared in multicolor ink or pencil, and on a variety of papers and films. Furthermore, the stylus can be equipped with a scriber for various scribing operations.

Digital plotting and drafting equipment operate on the three straight line axes of the cartesian coordinate system (see Figure 11-4). All topographic information is located on the X (latitude) and Y (longitude) axes. The third axis (Z) is used to raise and lower the drawing stylus.

Two broad categories of digital graphic drawing machines are available: flatbed plotters, and drum plotters.

Flatbed Plotters

As their name indicates, *flatbed plotters* consist of a flat bed or base upon which the drawing medium is supported, Figure 11-5. The stylus is mounted on a bridge which is capable of moving over the en-

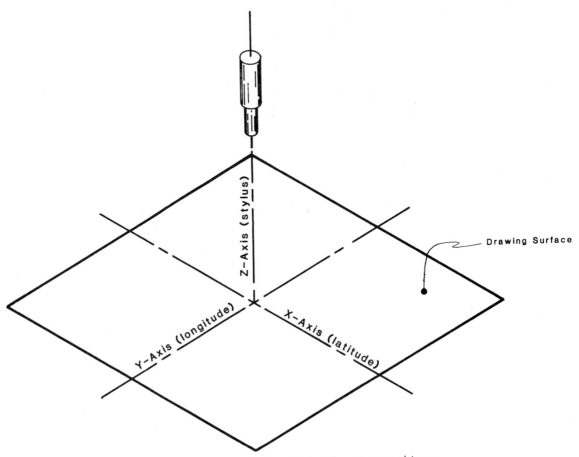

FIG. 11-4 Cartesian coordinate system applied to digital drawing machines

FIG. 11-5 Flatbed plotter (*Courtesy of Carl Zeiss, Inc., Photogrammetric Instruments Division*)

tire bed. (For multicolor drawings, a stylus is provided for each color, thereby making it possible to have several styluses on a bridge.) The stylus, in turn, moves up and down the bridge. Hence, the two movements correspond to the X and Y coordinates.

Plotting points and lines on the plotter can be accomplished in either of two ways. One procedure establishes a fixed point of origin (where the X and Y axes intersect). In Figure 11-6, the origin is in the center of the drawing, making it necessary to have both positive and negative X and Y values. This serves as a reference for all points on the drawing.

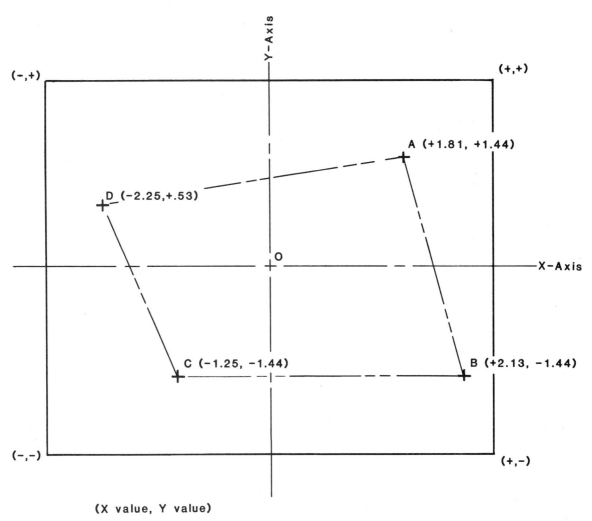

FIG. 11-6 Plotting points with fixed origin. All location measurements for each point are located in relationship to a fixed origin "O."

The second plotting procedure has no fixed point of origin nor reference. The location of a point is based upon its location relative to the last point located. This type of data is represented as ΔX and ΔY. Again, ΔX and ΔY can have either positive or negative values, Figure 11-7.

Drum Plotters

The second category of digital graphic drawing machines is the drum plotter. As can be seen in Figure 11-8, the bed of the plotter is cylindrical, and is referred to as a *drum*. The bridge is stationary, although the stylus can move along the bridge. To compensate for the stationary bridge, the drum rotates and thereby moves the paper back and forth. Thus, the X and Y coordinates correspond to the drum and stylus movement.

The programming used in drum plotters is the same as for flatbed plotters. The primary advantage of the drum plotter is that the drawing or paper can be continuous.

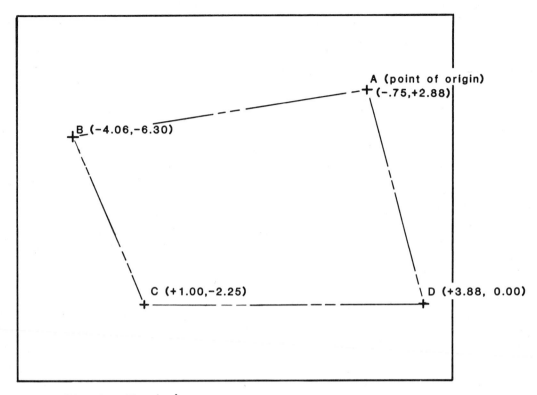

(X value, Y value)

FIG. 11-7 Plotting points with no fixed origin. All location measures are relative to previous position of stylus. In this example, stylus began at point *A*, and progressed to points *B*, *C*, and *D*. Second reading for *A* is positioned relative to *D*.

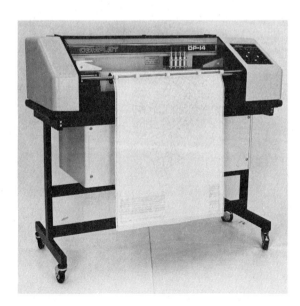

FIG. 11-8 Drum plotter (*Courtesy of Keuffel & Esser Company*)

While the flatbed drawing is confined to the dimensions of the bed (length and width), the drum plotter is confined only to the width of the drum; the length of the paper can be continuous.

11.3
INTERACTIVE COMPUTER GRAPHICS

Certain processes are performed better by a human being; others are better performed by machines. By combining the best of each, we frequently obtain a higher-quality product (output) at a lower cost (cost effectiveness).

One area that requires this essential combination is *interactive computer graphics*.

This field involves the combination of human and computer abilities in a direct and immediate communication linkage for design, production, and servicing activities. The availability of interactive computer graphics equipment, in turn, has stimulated activity by engineers and scientists in the field of *interactive problem solving.*

Interactive computer graphics is the continual and instant communication between a person and a computer graphics system. Interactive problem solving is continual and instant communication between a person and a computer graphics system for the purpose of solving a particular design, production, and/or servicing problem. In both situations, a visual display is required. This is different from using a digital graphics system in that the digital system does not allow for instantaneous two-way communications.

In interactive computer graphics, a person must set goals, alternatives, criteria, and direction, and is responsible for questioning and applying intuition to a problem. In short, the individual is responsible for identifying and making decisions about the unforeseen or unexpected.

Careful consideration must be given to the combined use of personnel and computer graphics equipment. Only by clearly understanding the roles and limitations of each can a well-designed interactive and integrated graphics system be designed, implemented, and effective. Interactive systems, then, require maximum utilization of both human and machine capabilities. This can only be accomplished by having clear and direct communications between personnel and computer.

Interactive Graphics Displays

The most familiar form of communication with a computer involves the use of punched cards for input and printed paper for output. This type of system requires completing a data form or sheet, punching cards, checking, correcting, submitting for computer run, and, finally, receiving the output. Because of the bulkiness and the time required for this system, it is gradually being replaced by the more rapid interactive computer graphics system.

Today there are numerous alternatives to the traditional punch-card format. Advances in interactive computer graphics and hardware have provided a choice of suitable systems and costs. Only a few years ago the costs were prohibitive, but now they are low enough to warrant serious consideration by the professional who generates and uses topographic drawings.

A computer graphical display system is, without a doubt, one of the most fascinating devices in computer technology ever pro-

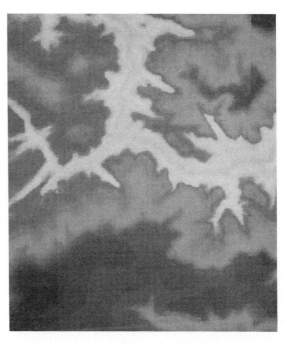

FIG. 11-9 A computer enhancement of satellite photogrammetric data and photograph (*Courtesy of Interactive Digital Image Manipulation Systems, a subsidiary of* TRW, Inc.)

duced. Many applications can take advantage of computer displays. For example, one can display satellite photogrammetric data and pictures of objects in space, and by rotating them gain visualization and enhancement into their total structure. Another use is *computer enhancement* (Figure 11-9), which emphasizes the features of objects to present a clearer presentation.

Interactive graphics displays are widely employed in engineering projects. Displays are used to design highway and drainage systems, presenting the design in perspective or layout. (See Figure 11-10.) By using the interactive system, the engineer can "play out" a variety of scenarios to determine the best design. With traditional methods, such scenario play-outs would be very time consuming as well as extremely costly. With the interactive mode, however, the cost is minimal and the turnaround time is almost instantaneous.

It should be mentioned that the display used in interactive computerized graphics is not considered hard copy. That is, it is not presented on a sheet of paper or film. It appears on a viewing screen. Two-way communication occurs via the use of a keyboard similar to that found on typewriters and the viewing screen. There is no need to keypunch cards or tapes, the system asks questions, and the operator gives the input.

Picture Production

Interactive computer graphics originated as an attempt to use the *cathode ray tube* (CRT) as an output device. The first use of a CRT output device was in 1950, with the development of Whirlwind I. Since then,

the CRT has maintained its status as virtually the only display method suitable for generating graphical output at high speeds. The advantages and limitations of the CRT have had a significant effect on the development of computerized graphics.

The principal disadvantage of the CRT is its inability to maintain a picture on the screen. This problem can be solved by refreshing the CRT from data stored in the computer's memory. However, this technique places a premium on the speed at which lines can be drawn, since the display will flicker if too many lines are being displayed. Special *direct-view storage tubes* have recently been developed which avoid the need for refreshing, but they are considered less versatile than conventional CRT units.

Another development that overcomes the problems associated with CRTs is the *plasma panel*. The viewing screen of the plasma panel consists of a series of tiny neon gas discharge cells (dots). These cells are arranged in a matrix of approximately 60 cells per lineal inch. The resulting image is similar to the half-tone picture typically found in the printed media. In addition to eliminating refresher tubes, the plasma panel is also significantly less expensive than CRT units.

In recent years, extremely powerful *display processors* have been built which are capable of applying a wide range of transformations (such as rotations, scaling, and distortions) to the display picture. Furthermore, progress has also been made in the production of inexpensive display processors and terminals. Low-cost terminals and time-sharing systems make possible wider application of interactive computer graphics.

Programming the Display

Interactive computer graphical programs are written by specifying a sequence of operations that produce a list of instructions for the display processor. The list of display instructions is called the *display file*. There is no standard way in which the display file must be constructed; the design is left to the ingenuity of the programmer.

A basic requirement of an interactive system is that it must be able to change the picture in a dynamic fashion. Often, an engineer would like to remove a part of the picture or change its positioning. This can be accomplished in one of two ways. One way is to reprogram the entire display file. Another way is to segment the display file and reprogram the part that is to be changed.

In order to simplify the use of repeated symbols or expressions in a picture, *display subroutines* are used. These are shared segments of display codes, somewhat analogous to conventional subroutines. The concept of using symbols is frequently extended to allow these symbols and expressions to be individually scaled and rotated.

A set of functions for building and manipulating a segmented display file is known as a *display file compiler.* These functions can generally be written without much difficulty. Problems do arise, however, when one wishes to specify the scaling and rotation of various parts of the picture. To do this it is necessary to have a simple notation for defining these transformations, and a means for detecting and removing parts of the picture. The first problem is solved by using matrices to define trans-formations, and the second involves the use of clipping programs.

The transformation of pictures with matrices represents an application of the techniques of algebra and trigonometry. *Clipping*, in comparison, is a technique found only in computer graphics. Various techniques, some based on hardware and others on software, have been proposed for performing transformations as rapidly as possible.

Interacting with the System

A number of devices, such as tablets, joysticks, and light pens have been developed for the input of graphical information into computer systems. When used in conjunction with a display, input devices make it possible to interact effectively with the system. The devices are used to draw lines and position symbols on the screen, or to point at items in order to change or delete them.

A number of ingenious techniques have been developed for using these devices as interaction tools. An example is *visual feedback*, which permits the use of relatively inaccurate input devices that need not write directly on the screen surface.

Programming input devices is a somewhat unconventional task, because they communicate with the computer system in unconventional ways. Furthermore, the user frequently makes use of two or more input devices at one time. Generally, the simplest way to handle inputs is by means of *interrupt routines*, which receive the input data and pass it on to the main program in the form of an *attention code*.

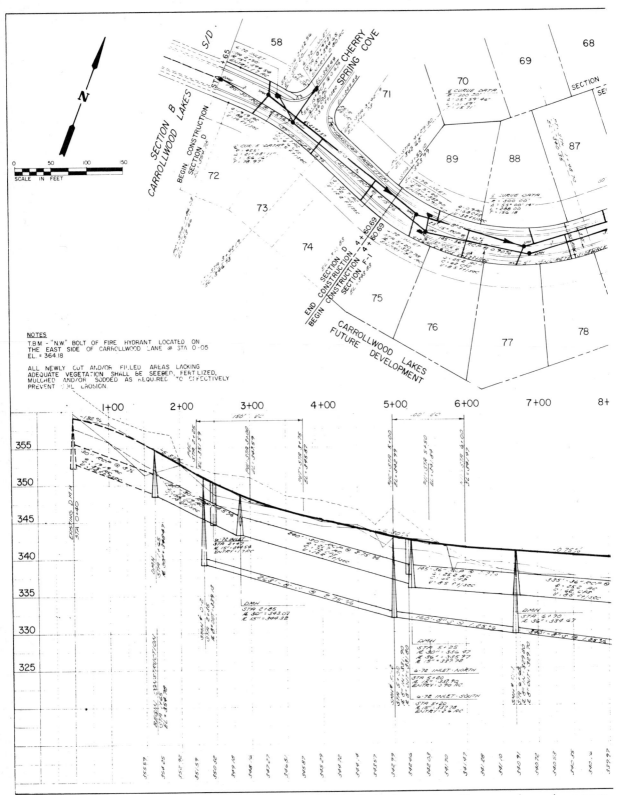

FIG. 11-10 Computer presentation for road design. Both plan and profile were drawn by a com-

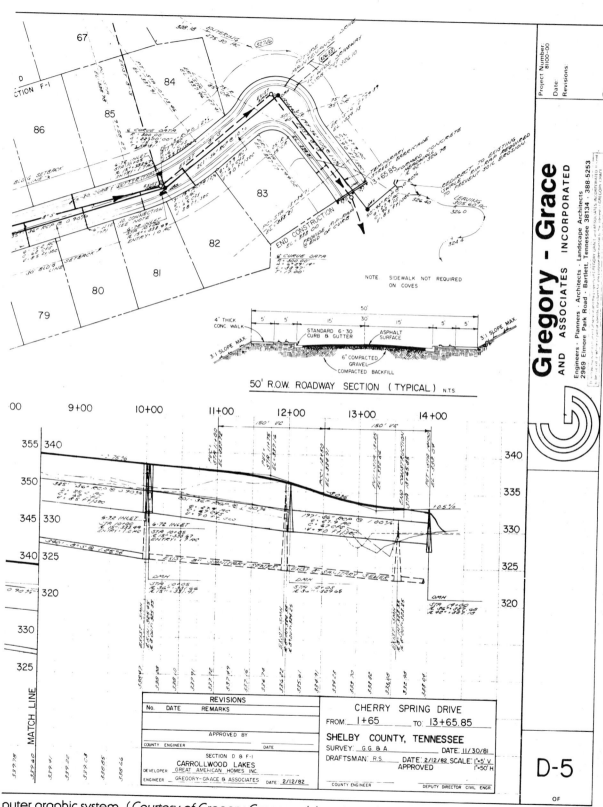

NOTE   SIDEWALK NOT REQUIRED
ON COVES

50' R.O.W. ROADWAY SECTION (TYPICAL) N.T.S.

STANDARD 6-30
CURB & GUTTER

ASPHALT
SURFACE

4" THICK
CONC WALK

3:1 SLOPE MAX

3:1 SLOPE MAX.

6" COMPACTED
GRAVEL
COMPACTED BACKFILL

REVISIONS		
No.	DATE	REMARKS

APPROVED BY

COUNTY ENGINEER   DATE

SECTION D & F-1
CARROLLWOOD LAKES
DEVELOPER   GREAT AMERICAN HOMES INC.
ENGINEER   GREGORY-GRACE & ASSOCIATES   DATE 2/12/82

CHERRY SPRING DRIVE
FROM: 1+65   TO: 13+65.85

SHELBY COUNTY, TENNESSEE
SURVEY: G.G & A   DATE: 11/30/81
DRAFTSMAN: R.S   DATE: 2/12/82 SCALE: 1"=5' V.
APPROVED   1"=50' H

COUNTY ENGINEER   DEPUTY DIRECTOR CIVIL ENGR.

D-5

OF

Gregory - Grace
AND ASSOCIATES INCORPORATED
Engineers · Planners · Architects · Landscape Architects
2969 Elmore Park Road · Bartlett, Tennessee 38134 · 388-5253

Project Number: 8100-00
Date:
Revisions:

puter graphic system. (*Courtesy of Gregory-Grace and Associates, Inc.*)

205

11.4 SUMMARY

Advancements in computer technology have increased the availability and use of computers by engineers, geographers, cartographers, and geophysicists. One area of computer technology is computer graphics, which includes the hardware and software used to generate graphical end products. Computer graphics are capable of generating line drawings and pictorial presentations.

Generating a graphic end product involves four processes. These are modeling, translation of data, data input, and data output.

There are two broad areas of computer graphics. The first and most familiar is digital graphics. This pertains to those systems that require input in digital form, such as punch cards and tapes. The two main categories of digital graphic drawing machines are flatbed plotters and drum plotters.

A second broad area of computer graphics is interactive computer graphics. This is any system that allows the continual and instant communication between a person and a computer system. Interactive computer graphics systems permit a person to communicate with a computer and see graphical output without delay, and they decrease the time and cost involved in digital systems. Interactive systems, however, do not provide instantaneous hard copies.

The primary medium used in interactive systems is the cathode ray tube, or CRT. Though it provides instant feedback and illustration, a major drawback of the CRT is its inability to hold a picture for any length of time. This can be overcome by refreshing the picture and using direct-view storage tubes. An alternative to CRTs is the plasma panel, which employs a matrix of neon gas discharge cells for image representation.

Programming interactive graphical systems involves specifying operations that produce a list of instructions for the display unit or processor. By using transformations, it is possible to manipulate and change the image shown on the CRT. Interacting with the system can be accomplished with input devices such as tablets, joysticks, and/or light pens.

KEY TERMS

Attention Code
Bridge
Cartesian Coordinate System
Clipping Program

Display Subroutine
Drum Plotter
Flatbed Plotter
Graphic Display

Output
Plasma Panel
Program
Programming
(*Continued*)

Computer Enhancement
Computer Language
Computer System
Computerized Graphics
CRT
Digital Graphics
Direct-view Storage Tube
Display File
Display File Compiler
Display Processor

Hardware
Input
Interactive Computer Graphics
Interactive Problem Solving
Interactive System
Interrupt Routine
Joystick
Light Pen
Modeling
Magnetic Tape

Punch Card
Punched Tape
Refreshing
Segment
Software
Stylus
Tablet
View Storage Tube

REVIEW

1. Explain computerized graphics, and give examples of how it can be used in the topographic drawing field.

2. Discuss the difference between computer hardware and software.

3. What application does modeling have in topographic work in relation to civil engineering projects?

4. Explain the purpose of a computer language, and describe its relationship to data input.

5. What is the difference between soft data output and hard data output?

6. How does digital graphics differ from interactive graphics?

7. Explain how digital graphics drawing machines function on the cartesian coordinate system.

8. What are the two ways in which points and lines are plotted on digital equipment?

9. Explain the relationship between interactive computer graphics and interactive problem solving.

10. What is the advantage of combining human talents and computer functions in topo projects?

11. Explain how interactive systems can be used to play out various engineering scenarios, as they apply to topographic concerns.

12. What is the difference between CRTs and plasma panels?

13. Why is refreshing necessary for CRTs?

14. Explain how removing a part of a picture is handled in computer graphics programming.

15. How can a person interact with a CRT?

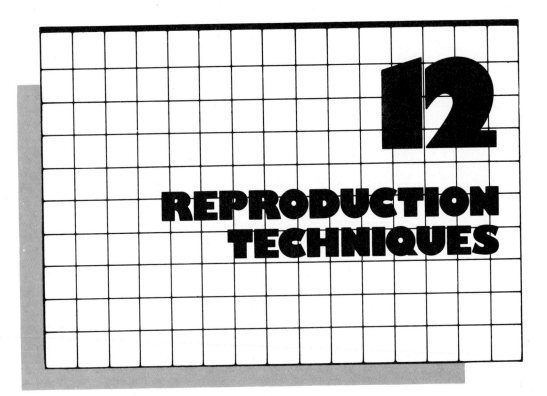

12
REPRODUCTION TECHNIQUES

The success of a topographic map draw- ing project can be determined by finding out if it accomplishes what it was designed to do. Does the drawing communicate all information clearly and accurately, and is it understood by the intended audience? If these questions are answered in the affirmative, then the drawing project is a success.

The last step in producing a topographic map product is the reproduction of the drawing. The reproduction technique se- lected can significantly influence the inter- pretation of the drawing. The reproduction technique can help in the clarification of vast quantities of information, as well as helping the readability of written data.

The reproduction technique, therefore, must be selected with great care. First, the audience and purpose of the reproduction must be determined. Topographic draw- ings used by contractors for a construction project, for example, will be reproduced in a significantly different way from topo- graphic maps used for military trans- portation planning.

12.1
CONSIDERATIONS FOR REPRODUCTION

Deciding on the type of reproduction technique involves consideration of several factors. The first factor is the *quantity* of reproductions required. If one needs 10 or fewer prints, a diazo reproduction may be more advantageous. If 10 to 50 prints are required, a photocopy or electrostatic (xerography) product may be desirable. Larger quantities may require consideration of photo-offset or engraving products. It should be noted that currently few maps are printed by the gravure process (from an incised or engraved image); these maps are found mostly in periodicals and some journals.

The second factor for consideration is the *quality* of reproduction required. Drawings to be used in engineering and architectural projects do not require high-quality reproductions of drawings. The diazo reproduction is commonly used for such cases. Where drawings are used for publication purposes (atlases) or for sales, however, more sophisticated reproduction techniques are used. These techniques include photographs (prints and slides), engraving prints, photo-offset prints, and multicolor xerography prints.

The third factor is *color*. Maps can be in single color or multiple colors. Single-color prints can be produced by all reproduction techniques. Multicolor reproductions, however, are limited to certain types of reproduction techniques, such as photo-offset, photographs, engraving prints, and color xerography.

The final factor for consideration is *cost*. How much money can be spent on the reproduction of the drawing? Naturally, the cost is affected by the first three factors: quantity, quality, and color. For example, using photo-offset would be an extremely expensive way to reproduce 5 multicolor prints. However, the technique would be cost efficient in reproducing 1000 multicolor prints.

12.2 ENGINEERING DRAWING REPRODUCTIONS

The majority of all engineering topographic drawings are made on tracing paper, vellum, or film, for the primary purpose of producing clear reproductions. Original or master drawings are of limited use to field survey teams and engineering projects if they cannot be reproduced and used in the field. Master topographic drawings are used almost entirely in the field and to create copies for annotation. Reproductions or prints (also called blueprints) are supplied to financial backers and lending agents, estimators, contractors, and others concerned with topographic features. The originals, or master drawings, are usually retained by the engineering firm that prepared them.

Several types of reproduction processes are commonly used in making engineering topo prints. Most require that the original be on translucent paper or film, and all require dark line work and lettering. These reproduction processes are blueprints, diazos, intermediates, xerographs, and photocopies.

Blueprints

At one time, the *blueprint* was the most common form of drawing reproduction, but is now used on a limited basis. The name is derived from the deep blue background of the print. Lines and lettering appear as white, and give the appearance of a negative print.

Blueprints are made by exposing the sensitized side of the paper, when in direct contact with the original, to ultraviolet light.

Once exposed, the sensitized paper is sent through a series of baths. The first is a fixing bath of potassium bichromate, and the second is a clear water rinse. Blueprinting is a *wet process;* therefore, the prints must be dried, which is a time-consuming process.

Blueprinting is unpopular due to several reasons. One reason is the time required to produce a finished print. A second reason is the shrinkage that occurs during the drying stage. A third reason is color deterioration when exposed to sunlight. There is also a relatively high cost per print, and the wet process causes dimensional instability, which is subject to scale changes. Because of these drawbacks, the blueprint has been replaced by the diazo print.

Diazos

Diazo prints are the most common means of reproducing engineering drawings. To produce usable prints, the drawings are prepared on transparent or translucent media. The prints appear as positive images; that is, there is a white background with dark lines. These lines may be blue, black, or sepia (brown) in color.

Diazo prints are also called *ozalid prints* or *ammonia vapor prints.* They are made by exposing the light-sensitive (diazo chemical sensitizers) coated side to ultra-violet light. The coated side is exposed while it is in direct contact with the original drawing. Once exposed, the sensitized paper is exposed to ammonia vapors which develop the print. The sensitized coatings are available on bond, tracing, vellum, and film surfaces. Diazo prints are made by a *dry process.*

The diazo print process has proven to be an economical form of reproduction. Prints can be reproduced fairly rapidly at a relatively low cost because they are produced via a dry process and have no shrinkage. The disadvantages are that diazos fade when exposed to sunlight, are not cost efficient for high-quantity printing, and do not produce a publishable copy.

Intermediates

Sometimes several masters are required. For example, there are cases where several contractors and subcontractors need to make special notations or specifications on the drawing and then reproduce it for their staff and work crews. To redraw or prepare a number of master or original drawings would be not only time consuming but also costly. To avoid these problems, *intermediate prints* are used.

Intermediates are also known as *sepias* and *Van Dykes.* They are reproductions or prints that function as a master or original drawing; that is, other reproductions can be made from them. The intermediate is made on translucent paper or film that has a light-sensitive coating, similar to diazo paper. The sensitized coating can be reproduced by either a wet or dry process, and produces prints with dark brown lines. For the dry process, the diazo sepia image is made on vellum or Mylar. The wet process is sometimes called Van Dyke or blueprinting. The Van Dyke is made from a negative. From this negative, either positive prints (dark brown prints) or positive blueprints (dark blue image) can be made.

When an intermediate is used in conjunction with the wet process of blueprinting, it serves as a negative. Blueprints made from intermediates have blue lines on a white background instead of white lines on a blue background. Dry process intermediates, on the other hand, always have a white background.

Xerographs

Xerographs are reproductions made by electrostatic copying. Originally used for the sole purpose of reproducing written materials, xerography is now used for reproducing line drawings as well as half-tone and continuous-tone illustrations. Xerographic systems have been advanced to the stage where they can reproduce, sort, and collate large numbers of original drawings in short periods of time.

When these systems were first available, they produced black line images on 8-1/2" × 11" or 8-1/2" × 14" sheets of paper. Today, xerographic systems can reproduce drawings at the size of the original or can reduce them down to any proportion. Some units can produce not only black line drawings, but also are capable of reproducing multicolor prints on various types of paper, film, and cloth.

Due to the dramatic and rapid advancements made in xerography, few firms have realized the full potential of these systems. Most engineering firms still use xerography as a means for reproducing written materials and reducing drawings down to standard-size paper. It is expected, however, that xerography will some day become the most common technique used to reproduce engineering drawings.

Photography

Engineering firms occasionally find that high-quality reproductions are necessary. In these situations they use *photocopy prints*. These reproductions are made on light-sensitive, high-contrast photographic paper, producing a photographic image of the drawing. Because of the high contrasting quality of photocopies, some pencil tones are difficult to define. Inked drawings, on the other hand, produce exceptionally good prints.

The use of photography allows the scale of the drawing to be changed without redrawing. This is accomplished by print reduction or enlargement. Many firms produce a positive photographic image on Mylar for their clients. Such prints can then be used to run inexpensive diazo prints. Also, images produced by aerial photographs can be delivered as half-tone or high-contrast images on Mylar, from which the client can produce other copies.

12.3
MAP REPRODUCTIONS

Reproduction techniques used for maps such as those published by the United States Geological Survey are significantly different from those found in engineering topo drawings. They require more sophisticated methods of reproduction, because maps must usually be reproduced in greater quantities and with better quality.

Two techniques used in map reproduction are *photo-engraving* and *photo-offset*. Both methods are capable of producing high-quality prints in large numbers. They employ photographic procedures that ensure the reproduction of mapping details and enable the production of multicolor products. The key to these two reproduction techniques, as their names imply, is the application of photography, particularly to multicolor maps.

Color Separation

Color map printing is accomplished by printing each color separately with ink. To accomplish this, the individual map colors must be separated so that an individual plate can be prepared to print each color. For single-color maps, only one photographic image and plate is needed; there is no need for color separation. Color separation can be accomplished by either of two methods: mechanical separation, or photographic separation.

It should be noted that single-color maps can be reproduced from a single piece of artwork. However, in larger publishing houses, several plates are used. One plate is used for type, one for line work, and another for one or more elements to be screened. From these plates, a composite negative and a single press plate are produced.

Mechanical separation is the method used most commonly in map reproduction. This technique is specifically set up for reproducing color line work, as is used in map drawings. For all practical purposes, mechanical separation is a form of overlay drawing, and uses overlay drawing procedures.

Mechanical separation requires that a separate drawing be prepared for each color on the map. This includes all notations, titles, names, and scales. Drawings are usually prepared on clear acetate overlays in ink instead of on paper or polyester film. The acetate sheets are hinged by tape on the drawing surface in proper alignment with the base drawing, which is usually the black image sheet. Once the acetate overlay drawings are completed, they are photographed separately. Each negative is then used to produce the respective color image on the reproduction plate.

The process used in governmental agencies and large publishing houses varies from the procedure just described. Each color drawing is produced on stable-base material (Mylar) by scribing line elements. Punch registration markers are used for alignment purposes. Each map feature is prepared on a separate layer (sheet of material) to help in any required revision work. These layers are then combined into a composite negative for platemaking.

Photographic separation is used only for color photographs such as aerial photos. Unlike the mechanical separation process, this technique requires no drawing. By means of color filters placed over the camera lens, a separate negative is developed for each color.

When the mechanical separation process might consist of numerous overlays, photographic separation is made up of four different negatives. In combination, they can produce any color desired. The four color separations are black, yellow, cyan, and magenta. The "color" image negative is screened (made up of a series of dots, with each color having its own screen angle or dot alignment). Again, the negatives are used to make the press plate for each printing ink.

Photo-engraving

Photo-engraving, or *gravure printing*, involves several steps prior to the actual printing process. This printing technique should be used only where there is need for an extremely large number of map reproductions. Because of the relatively complicated and time-consuming process of platemaking, it is cost efficient only where there is need for long runs on the printing press.

At one time, photo-engraving was the standard of the industry, but it has been replaced by photo-offset. Photo-engraving is a process commonly used to reproduce line drawings or half-tone photographs. This process, however, has limited use for map printing today. The only exceptions are in books and periodicals. The gravure process has taken a back seat to less expensive methods.

Photo-offset

Photo-offset printing is also called *photo-offset lithography*. It is a reproduction process in which water is used to keep the ink off nonimage areas of the plate. This process is the standard of the industry for high-quantity prints. Photo-offset is based upon the principle that water and grease do not mix. The greasy printing ink does not adhere to the nonimage portion of the plate, which is moistened with water prior to each inking.

The plate, usually of aluminum or zinc, can be made by either indirect platemaking or direct platemaking.

Indirect platemaking is the traditional method for making lithography plates. The film negative is first mounted on a layout sheet called *masking paper.* The paper is cut away from image areas on the negative. The remaining paper serves as a blocking or masking device during light exposure. Any scratches or pin holes in the negative are touched up with an opaquing paint.

Once the negative is mounted, the masking paper is placed in direct contact with the light-sensitive side of the plate, and it is then exposed to ultraviolet light. After exposure, the plate is developed in a sequence of three washes. The first wash is in *desensitizing gum,* which removes sensitizing coating from all areas of the plate not exposed to light. The second is in *developing lacquer;* the lacquer serves as a fixing chemical in that it makes the image permanent. Lastly, gum arabic is used to extend the life of the lithography plate.

Direct platemaking is so named because the image from the master drawing is made directly onto the plate itself, with no need of a film negative. In place of the film, a specially treated plate is mounted in the camera itself and exposed to the drawing. The plate is then developed in a process similar to film development. In some direct-platemaking units, it is possible to expose the plate and develop it without human contact. Recent advancements have also eliminated the need for the traditional camera, using *laser scanners* for image impressions. The lasers provide a film image, which is used to produce the press plate.

Once the lithographic plate is made, it is mounted on the master cylinder of an offset press. While the plate is rotating on the master cylinder, ink and water are applied to the plate. The master cylinder is then brought into contact with the blanket cylinder; this cylinder has a rubberized blanket that picks up the ink from the master cylinder. The ink is transferred to the paper as it is fed between the blanket and impression cylinders; hence the name *offset.* See Figure 12-1. This process is repeated for each ink.

It should be noted that a few large presses are capable of printing all four inks

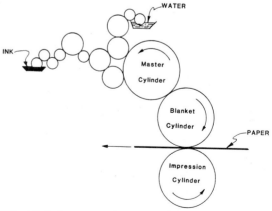

FIG. 12-1 The photo-offset printing process

on the same run. These presses contain four master cylinders, one for each ink. As the paper is printed in one color, it is automatically passed on to the next, until all four inks are printed.

12.4
DRAWING STORAGE

Legal requirements and company practices often dictate that original drawings be stored and kept available for specific periods of time.

Topographic drawings may be stored as part of a set of working drawings, or as a separate set unto themselves. The first practice is commonly found in the engineering and architectural fields. The second is more typical in cartography and mapping.

The method selected for storage depends upon what is being stored and for what period of time. Drawings stored for short periods of time, such as for the duration of a project, are stored in *drawers* or *drawing racks*. Drawers used for drawing storage should be large enough to hold a drawing so that it can lay flat without being folded or rolled. Such drawers are typically designed to hold a depth of anywhere from 1 inch to 3 inches of drawings.

Rack storages are made of wood or metal. Hinges are attached to the top or side of the drawings so that they can be opened like a book. The hinges can then be used to hang the drawings on racks which are either permanently fixed to walls or which are portable units. The purpose of drawer and rack storage is to make the drawings accessible to users, and to reduce the amount of storage space.

For longer storage times, usually from 7 to 10 years, *storage vaults* are used. Before the drawings are placed in the vaults, they are rolled up and put into *storage tubes.* These tubes are made of cardboard, plastic, or metal. In turn, the tubes are stored in the vaults on racks. In some cases, the vaults' temperature and humidity are controlled. Such systems can be expensive, especially when there are large numbers of drawings that must be stored. For map materials involving multiple materials, flat storage is preferred. Also, some layers might include adhesive type or sheets that do not store well in rolled form.

Some drawings need not be stored in their original form. In these cases, a more cost-effective storage method is used: *photographic reduction.* As its name implies, this method involves the photographic reduction of original drawings down to standard paper sizes (8-$1/2''$ × 11" or 8-$1/2''$ × 14") or smaller. Drawings are frequently reduced down to one-third their original size on process cameras, and even smaller on microfilm cameras.

Photographic reductions can be stored in standard-size folders, index cards, and film reels. Hence, the amount of storage space is radically reduced from that required for traditional storage methods. It is anticipated that in the future more drawings will be photographically reduced and stored.

12.5 SUMMARY

An important part of preparing a topographic drawing is reproduction. Since there are various reproduction techniques available, care should be taken in selecting the most appropriate procedure. Factors that should be taken into consideration are the quantity of reproductions needed, the quality of reproduction expected, whether or not multicolor reproduction is essential, and how much money should be spent on reproduction.

Engineering reproductions are different from map reproductions. They are usually of lower quality but are faster and less expensive. The techniques frequently used in engineering are blueprints, diazos, intermediates, xerographs, and photocopies.

Map reproductions are usually of such quality that they are publishable. They also are frequently reproduced in multicolors and in large quantities. Before printing, color maps must be prepared so that the colors are separated into separate drawings. This is accomplished by either mechanical or photographic separation techniques. Once color separation is accomplished, the maps can be reproduced by either photo-engraving or photo-offset methods.

Due to legal and in-house requirements, original drawings must be kept for specific periods of time. For short periods of time, drawer storage and rack storage methods are used. For longer periods, it is necessary to use storage vaults or photographic-reduction storage.

KEY TERMS

Ammonia Vapor Print
Blanket Cylinder
Blueprint
Carbon Tissue
Color
Color Separation
Cost
Desensitizing Gum
Developing Lacquer
Diazo
Direct Platemaking

Gum Arabic
Impression Cylinder
Indirect Platemaking
Intermediate Print
Laser Scanner
Lithography
Masking Paper
Master Cylinder
Master Drawing
Mechanical Separation
Negative

Photographic Reduction
Photographic Separation
Photo-offset
Print
Quality
Quantity
Reproduction
Sensitized
Sepia
Storage Tube
Storage Vault

(Continued)

Drawing Rack Storage
Drawer Storage
Dry Process
Gravure

Opaquing Paint
Ozalid
Photocopy
Photo-engraving

Van Dyke
Wet Process
Xerography

REVIEW

1. Explain why it is important to select the right type of reproduction technique.

2. Identify and explain the significance of the four factors that must be taken into consideration when selecting the appropriate reproduction technique.

3. What are blueprints?

4. Describe diazo prints, and explain how they differ from blueprints.

5. Why is the diazo print the basic technique used in engineering drawing reproduction?

6. Explain the uses of intermediates.

7. What are xerographs?

8. In what ways is engineering drawing reproduction different from map reproduction?

9. Explain color separation, and describe the two methods used for this process.

10. Explain gravure printing and offset printing.

11. Discuss the different techniques of storing drawings over short periods of time and over long periods of time.

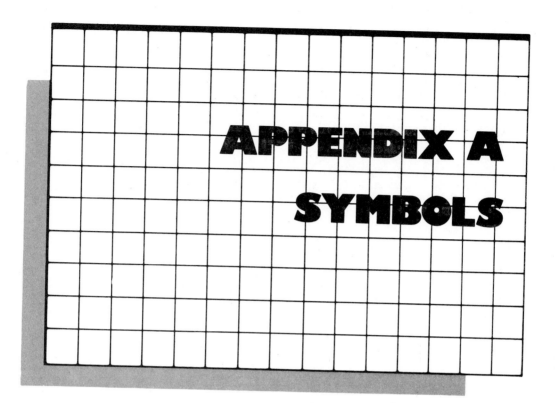

APPENDIX A
SYMBOLS

---

a	Semimajor axis of the earth ellipsoid.
A	Matrix of the coefficients of parameters in a least squares solution.
abc	Image space plane containing points a, b, and c.
ABC	Object space plane containing points A, B, and C.
a, b, c, etc.	Sometimes used instead of numbers to identify image points corresponding to the ground points A, B, C, etc.
A, B, C, etc.	Sometimes used instead of numbers to identify successive ground points or camera stations.
A^T	Transposed coefficient matrix.
A^{-1}	Inverse coefficient matrix.
b	Base length at the scale of the stereomodel; Coefficient of a parameter (*i.e.*, an element of B); semiminor axis of the earth ellipsoid.
B	Base length (*i.e.*, the physical distance between camera stations).

---

*The materials reproduced in this Appendix are used with permission from American Society of Photogrammetry, "List of Photogrammetry Symbols," *Manual of Photogrammetry*, American Society of Photogrammetry, 1980.

$\dot{B}$	Partial derivatives within observation equation matrix from exposure station parameters (partitioned matrix).
$\ddot{B}$	Partial derivatives within observation equation matrix from object point parameters (partitioned matrix).
b'	Image distance between principal point and corresponding (conjugate) principal point on the right hand photograph.
$b_x,\ b_y,\ b_z$	Components of the base length at the scale of the stereomodel.
$B_x,\ B_y,\ B_z$	Components of the base length in an exterior coordinate system.
c	As a superscript or subscript, used to identify the coordinates of a camera station (e.g., $X^c,\ Y^c,\ Z^c$, or the letters L or O also may sometimes be used; Focal length, often called camera constant.
C	Grid convergence; Degrees centigrade.
d	Displacement of a photographic image due to any cause, Diaphragm.
D	Determinant; Density.
$dh,\ dp$	Difference in height or elevation, difference in parallax; d is used here to imply difference, not differential.
e	Eccentricity of the earth ellipsoid.
E	Endlap, (overlap) between successive exposures along the flight line; exposure; easting in ground or grid coordinates.
f	Flattening of the earth ellipsoid; focal distance in optics; Alternative for (c).
F	Degrees Farenheit.
h	Height or elevation of a ground station or object above sea level datum (unless specified otherwise).
H	Height or elevation of camera stations above sea level datum (unless specified otherwise); alternative for (Z^o).
$h'h'$	Photograph, horizon.
i	Isocenter on a photograph; As a subscript, identifies the coordinates of the ith camera station (e.g., $X_i^c,\ Y_i^c,\ Z_i^c$); image distance in optics.
$\mathbf{i}$	Unit vector in direction of x-axis.
I	Apparent ground position of the isocenter.
$\mathbf{I}$	Unit or identity matrix.
j	As a subscript, identifies the coordinates of the jth ground point (e.g., $X_j,\ Y_j,\ Z_j$).
$\mathbf{j}$	Unit vector in direction of y-axis.
$\mathbf{J}$	Jacobian matrix.
$\mathbf{k}$	Unit vector in direction of z-axis.
K	Degrees Kelvin.
L	Lens point or camera station; may be used as a superscript or subscript to identify the coordinates of a camera station (e.g., $X_L,\ Y_L,\ Z_L$).
L_1	Left image exposure station.
L_2	Right image exposure station.
m	Scale number; alternative for standard error or standard deviation, (see also σ); metre.
M	Magnification; A factor which denotes the ratio of a dimension on a photographic copy to the corresponding dimension on the negative.
$\mathbf{M}$	A 3-by-3 orthogonal matrix of direction cosines, usually specifying the angular orientation between the photograph coordinate system and an exterior coordinate system.
$m_{11},\ m_{12},$ etc.	Elements of $\mathbf{M}$.
m_b	Photograph scale number.
m_k	Map scale number.
m_m	Model scale number.
n	Nadir point on a photograph; Index of refraction; number of measurements or observations.

N	Ground nadir point; Northing in grid coordinates; Radius of curvature for the prime vertical of the earth.
N	Matrix of the coefficients of a set of normal equations.
$\dot{N}$	Coefficient matrix of image exterior orientation elements (partitioned matrix).
$\ddot{N}$	Coefficient matrix of passpoint object space coordinates (partitioned matrix).
o	Sometimes used to designate a photograph principal point or zero positions; Object distance in optics.
O	Origin point of perspective, lens point, camera station; Origin of a system of space coordinates relative to a photograph.
p	Photograph principal point.
P	Ground principal point.
P	Alternative for weight matrix (inverse of covariance).
P_x	x-parallax corresponding to elevation difference.
P_y	y-parallax.
P_z	Vertical parallax in terrestrial stereomodel.
Q	Covariance matrix; Matrix of cofactors or weight coefficients.
r	Radial distance from any specified photograph center to an image; Redundancy, or degrees of freedom.
s	Swing angle.
S	Sidelap between adjacent flight strips; scale (either fractional/ratio/decimal representation).
t	Tilt of a camera axis from the vertical; Time.
T	As a superscript, used to indicate the transpose of a matrix (e.g., M^T); Transmission.
t_x	x-tilt, lateral tilt; Angular component of the resultant tilt, measured about the x-axis.
t_y	y-tilt, longitudinal tilt; Angular component of the resultant tilt, measured about the y-axis.
u	Number of unknowns.
v	Speed; The coefficient of a residual, (V).
λ, μ, ν	Direction cosines (lambda, mu, nu).
μm	Micrometre or micron, 10^{-6} metre.
ρ (rho)	Conversion factor from radian to angular measure (e.g., $\rho'' = 206,264,8''$).
σ (sigma)	Standard deviation or standard error.
σ^2	Variance.
$\dot{\sigma}$	Least squares correction vectors to exposure station parameters (partitioned matrix).
$\ddot{\sigma}$	Least squares correction vectors to object point coordinates (partitioned matrix).
ϕ (phi)	Pitch, longitudinal tilt, tilt about the y or Y axis; Latitude.
Φ (PHI)	Model rotation about Y-axis.
ω (omega)	Roll, lateral tilt, tilt about the x or X axis; In missile photogrammetry, elevation angle; Rotational velocity of the earth.
Ω (OMEGA)	Model rotation about X-axis.

SYMBOLS AND CONVENTIONAL SIGNS OF GREAT BRITAIN

BRACKEN

CONIFEROUS TREES
(NOT SURVEYED)

DECIDUOUS TREES
(NOT SURVEYED)

MARSH, SALTINGS

REEDS

ANTIQUITY SITE

FBM   BENCH MARK
(FUNDAMENTAL)

•ts   PERMANENT
TRANSVERSE STATION

SURFACE LEVEL

BUSH

COPPICE

FURZE

ORCHARD TREES

ROUGH GRASSLAND

WATER FLOW DIRECTION

CAVE ENTRANCE

•rp   REVISION PT.
(INSTRUMENTALLY FIXED)

TRIANGULATION STATION

PERIMETER OF BUILT-UP
AREA WITH SINGLE
ACREAGE
(1:2500 SCALE ONLY)

CONIFEROUS TREES
(SURVEYED)

DECIDUOUS TREES
(SURVEYED)

HEATH

OSIERS

SCRUB

BM   BENCH MARK (NORMAL)

ELECTRICITY PYLON

rp   REVISION PT. AND
BENCH MARK

AREA BRACE
(1:2500 SCALE ONLY)

ROOFED BUILDINGS		INACTIVE QUARRY CHALK PIT
BOULDERS		COUNTY BOUNDARY (GEOGRAPHICAL)
CULVERT		COUNTY AND CIVIL PARISH BOUNDARY COTERMINOUS
		ADMINISTRATIVE COUNTY OR COUNTY BOROUGH BOUNDARY
ACTIVE QUARRY CHALK OR CLAY PIT		LONDON BOROUGH BOUNDARY
DUNES		CIVIL PARISH BOUNDARY
SAND		BORO (OR BURGH) CONST & WARD BOUNDARY

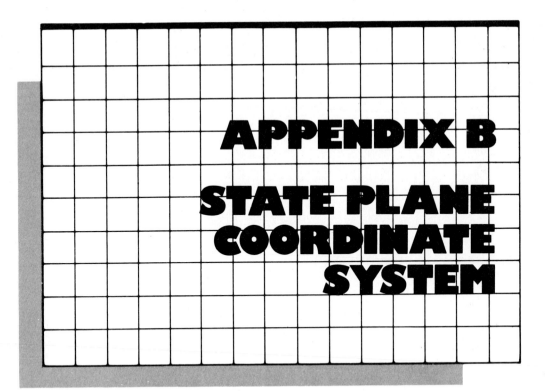

APPENDIX B
STATE PLANE COORDINATE SYSTEM

The mapping grid systems used in the United States vary from state to state, and, in some cases, from state zone to state zone. The following is a listing of the states, their zones, and the grid system used per zone. It will be observed that there are two grid systems used in the United States: the Lambert conformal projection, and the transverse Mercator projection.

State and Zone	Mapping Grid System

Alabama
    eastern
    western
	transverse Mercator projection
	transverse Mercator projection

Alaska
    zone 1
    zones 2–9
    zone 10
	oblique transverse Mercator projection
	transverse Mercator projection
	Lambert conformal projection

State and Zone	Mapping Grid System

Arizona
  eastern
  central
  western
	transverse Mercator projection
	transverse Mercator projection
	transverse Mercator projection

Arkansas
  northern
  southern
	Lambert conformal projection
	Lambert conformal projection

California zones 1–7	Lambert conformal projection

Colorado
  northern
  central
  southern
	Lambert conformal projection
	Lambert conformal projection
	Lambert conformal projection

Connecticut	Lambert conformal projection
Delaware	transverse Mercator projection

Florida
  eastern
  western
  northern
	transverse Mercator projection
	transverse Mercator projection
	Lambert conformal projection

Georgia
  eastern
  western
	transverse Mercator projection
	transverse Mercator projection

Hawaii zones 1–5	transverse Mercator projection

Idaho
  eastern
  central
  western
	transverse Mercator projection
	transverse Mercator projection
	transverse Mercator projection

State and Zone	Mapping Grid System

Illinois
  eastern
  western
	transverse Mercator projection
	transverse Mercator projection

Indiana
  eastern
  western
	transverse Mercator projection
	transverse Mercator projection

Iowa
  northern
  southern
	Lambert conformal projection
	Lambert conformal projection

Kansas
  northern
  southern
	Lambert conformal projection
	Lambert conformal projection

Kentucky
  northern
  southern
	Lambert conformal projection
	Lambert conformal projection

Louisiana
  northern
  southern
	Lambert conformal projection
	Lambert conformal projection

Maine
  eastern
  western
	transverse Mercator projection
	transverse Mercator projection

Maryland	Lambert conformal projection

Massachusetts
  mainland
  island
	Lambert conformal projection
	Lambert conformal projection

Michigan[1]
  eastern
  central
  western
	transverse Mercator projection
	transverse Mercator projection
	transverse Mercator projection

[1]Computed at an elevation of 800 feet above sea level, rather than at sea level

State and Zone	Mapping Grid System

Minnesota
  northern
  central
  southern
	Lambert conformal projection
	Lambert conformal projection
	Lambert conformal projection

Mississippi
  eastern
  western
	transverse Mercator projection
	transverse Mercator projection

Missouri
  eastern
  central
  western
	transverse Mercator projection
	transverse Mercator projection
	transverse Mercator projection

Montana
  northern
  central
  southern
	Lambert conformal projection
	Lambert conformal projection
	Lambert conformal projection

Nebraska
  northern
  southern
	Lambert conformal projection
	Lambert conformal projection

Nevada
  eastern
  central
  western
	transverse Mercator projection
	transverse Mercator projection
	transverse Mercator projection

New Hampshire	transverse Mercator projection
New Jersey	transverse Mercator projection

New Mexico
  eastern
  central
  western
	transverse Mercator projection
	transverse Mercator projection
	transverse Mercator projection

State and Zone	Mapping Grid System

New York
  Long Island
  eastern
  central
  western
	Lambert conformal projection
	transverse Mercator projection
	transverse Mercator projection
	transverse Mercator projection

North Carolina	Lambert conformal projection

North Dakota
  northern
  southern
	Lambert conformal projection
	Lambert conformal projection

Ohio
  northern
  southern
	Lambert conformal projection
	Lambert conformal projection

Oklahoma
  northern
  southern
	Lambert conformal projection
	Lambert conformal projection

Oregon
  northern
  southern
	Lambert conformal projection
	Lambert conformal projection

Pennsylvania
  northern
  southern
	Lambert conformal projection
	Lambert conformal projection

Rhode Island	transverse Mercator projection

South Carolina
  northern
  southern
	Lambert conformal projection
	Lambert conformal projection

South Dakota
  northern
  southern
	Lambert conformal projection
	Lambert conformal projection

State and Zone	Mapping Grid System
Tennessee	Lambert conformal projection

Texas
   northern
   north-central
   central
   south-central
   southern
	
	Lambert conformal projection
	Lambert conformal projection
	Lambert conformal projection
	Lambert conformal projection
	Lambert conformal projection

Utah
   northern
   central
   southern
	
	Lambert conformal projection
	Lambert conformal projection
	Lambert conformal projection

Vermont	transverse Mercator projection

Virginia
   northern
   southern
	
	Lambert conformal projection
	Lambert conformal projection

Washington
   northern
   southern
	
	Lambert conformal projection
	Lambert conformal projection

West Virginia
   northern
   southern
	
	Lambert conformal projection
	Lambert conformal projection

Wisconsin
   northern
   central
   southern
	
	Lambert conformal projection
	Lambert conformal projection
	Lambert conformal projection

Wyoming zones 1–4	 transverse Mercator projection

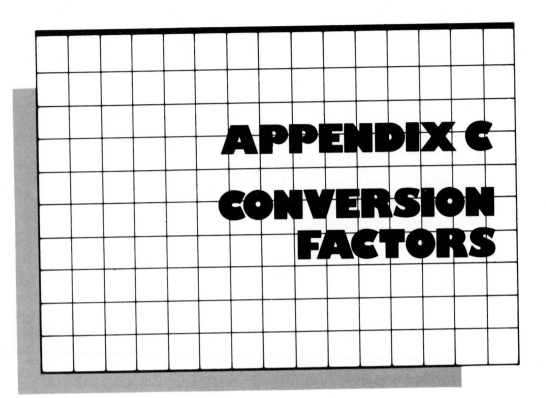

APPENDIX C
CONVERSION FACTORS

SURVEYOR'S (GUNTER'S) MEASUREMENT

1 link = 7.92 inches
1 inch = 0.12626 link
1 link = 0.66 foot
1 foot = 1.51515 links
1 rod = 25 links
1 link = 0.04 rod
1 rod = 16.5 feet
1 foot = 0.06061 rod

1 chain = 100 links
1 link = 0.01 chain
1 chain = 66 feet
1 foot = 0.015152 chain
1 chain = 4 rods
1 rod = 0.25 chain
1 mile = 80 chains
1 chain = 0.0125 mile

THE AMERICAN SURVEY FOOT

1 American Survey foot = 0.3048006 meter
1 meter = 3.28083333 feet
1 meter = 39.37 inches

THE FOOT SYSTEM

1 yard = 0.9144 meter	1 meter = 3.28083990– feet
1 foot = 0.348 meter (exactly)	1 meter = 39.370787+ inches

U.S. SYSTEM

1 foot = 12 inches	1 mile = 1760 yards
1 inch = 0.08333 foot	1 yard = 0.00056818 mile, statute
1 yard = 3 feet	1 mile, nautical = 6,076.1155 feet
1 foot = 0.33333 yard	1 foot = 0.00016458 mile, nautical
1 mile = 63,360 inches	1 mile, nautical = 2,025.3718 yards
1 mile = 5280 feet	1 mile, nautical = 1.150776 miles, statute
1 foot = 0.00018939 mile	1 mile, statute = 0.86898 mile, nautical

SURVEYOR'S (GUNTER'S) AREA MEASUREMENT

1 acre = 10 square chains	1 square yard = 0.0002066 acre
1 square chain = 0.10 acre	1 square mile = 6400 square chains
1 acre = 43,560 square feet	1 square chain = 0.00015625 square mile
1 square foot = 0.00002294 acre	1 square mile = 640 acres
1 acre = 4840 square yards	1 acre = 0.0015625 square mile

THE METRIC SYSTEM

Metric Linear Measurement

1 millimeter = 0.03937 inch	1 inch = 25.4 millimeters
1 centimeter = 0.39370 inch	1 inch = 2.54 centimeters
1 meter = 3.28084 feet	1 foot = 0.3048 meter
1 meter = 1.09361 yards	1 yard = 0.9144 meter
1 kilometer = 0.62137 mile, statute	1 mile = 1.609344 kilometer

Metric Area Measurement

1 square meter = 1 centiare
1 are = 100 square meters (centiares)
1 hectare = 100 ares
1 square kilometer = 100 hectares

1 square inch = 645.16 square millimeters
1 square millimeter = 0.00155 square inch
1 square inch = 6.4516 square centimeters
1 square centimeter = 0.155 square inch
1 square foot = 0.09290 square meter
1 square meter = 10.76426 square feet
1 square yard = 0.83613 square meter
1 square meter = 1.19599 square yard
1 acre = 0.40469 hectare
1 hectare = 2.47105 acres
1 square mile = 258.99881 hectares
1 hectare = 0.003861 square mile
1 square mile = 2.58999 square kilometers
1 square kilometer = 0.3861 square mile

THE VARA

The *vara* was a Spanish unit of linear measurement. Hence, in those states originally settled by Spain, the vara is sometimes encountered in deed searches and original deeds. Furthermore, the vara is also a Mexican unit of linear measurement. To avoid complications, many states have legally defined the equivalent measurements of the vara; they are presented here.

Mexico and Southwestern United States
1 vara = 32.99312 inches
1 vara = 4.1658 links
1 vara = 2.74943 feet
1 foot = 0.36371 vara

California
1 vara = 33 inches
1 vara = 4.1667 links
1 vara = 2.75 feet
1 foot = 0.36364 vara

Florida
1 vara = 33.372 inches
1 vara = 4.2136 links
1 vara = 2.7810 feet
1 foot = 0.35958 vara

Texas
1 vara = 33.333 inches
1 vara = 4.2088 links
1 vara = 2.77778 feet
36 varas = 100 feet
36 varas = 1.5152 chains
1900.8 varas = 1 mile, statute

THE ARPENT

The *arpent (arpen)* was a French unit of area measurement. The arpent was used in the early French settlements of North America. As in the case of the Spanish vara, the arpent has been legally defined in those states encountering the French area measurement. The states and their legal definitions of the arpent are listed here.

Arkansas and Missouri

1 arpent = 0.8507 acre

one side of a square arpent = 192.500 feet

one side of a square arpent = 2.91667 chains

Louisiana, Mississippi, Alabama, and Northwestern Florida

1 arpent = 0.84625 acre

one side of a square arpent = 191.994 feet

one side of a square arpent = 2.909 chains

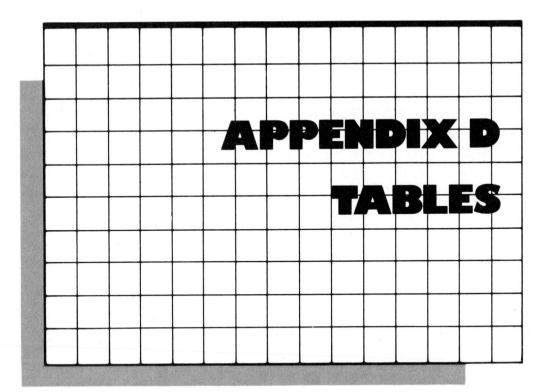

APPENDIX D

TABLES

In the appendix material on the following pages, you will find common trigonometric tables useful in the analysis of topographic values: Table A-1: Values of Trigonometric Functions and of Radians, Table A-2: Functions for Angles Greater than 90°; Table A-3: Powers, Roots, and Reciprocals 1 through 100; Table A-4: Constants; Table A-5: Lengths of Circular Arcs. Radius = 1; and Table A-6: Natural Versed Sines.

TABLE A-1: VALUES OF TRIGONOMETRIC FUNCTIONS AND OF RADIANS

Degrees	Radians	Sin	Csc	Tan	Cot	Sec	Cos			
0° 0'	0.0000	0.0000	—	0.0000	—	1.000	1.0000	1.5708	90°	0'
10'	029	029	343.8	029	343.8	000	000	679		50'
20'	058	058	171.9	058	171.9	000	000	650		40'
30'	0.0087	0.0087	114.6	0.0087	114.6	1.000	1.0000	1.5621		30'
40'	116	116	85.95	116	85.94	000	0.9999	592		20'
50'	145	145	68.76	145	68.75	000	999	563		10'
1° 0'	0.0175	0.0175	57.30	0.0175	57.29	1.000	0.9998	1.5533	89°	0'
10'	204	204	49.11	204	49.10	000	998	504		50'
20'	233	233	42.98	233	42.96	000	997	475		40'
30'	0.0262	0.0262	38.20	0.0262	38.19	1.000	0.9997	1.5446		30'
40'	291	291	34.38	291	34.37	000	996	417		20'
50'	320	320	31.26	320	31.24	001	995	388		10'
2° 0'	0.0349	0.0349	28.65	0.0349	28.64	1.001	0.9994	1.5359	88°	0'
10'	378	378	26.45	378	26.43	001	993	330		50'
20'	407	407	24.56	407	24.54	001	992	301		40'
30'	0.0436	0.0436	22.93	0.0437	22.90	1.001	0.9990	1.5272		30'
40'	465	465	21.49	466	21.47	001	989	243		20'
50'	495	494	20.23	495	20.21	001	988	213		10'
3° 0'	0.0524	0.0523	19.11	0.0524	19.08	1.001	0.9986	1.5184	87°	0'
10'	553	552	18.10	553	18.07	002	985	155		50'
20'	582	581	17.20	582	17.17	002	983	126		40'
30'	0.0611	0.0610	16.38	0.0612	16.35	1.002	0.9981	1.5097		30'
40'	640	640	15.64	641	15.60	002	980	068		20'
50'	669	669	14.96	670	14.92	002	978	039		10'
4° 0'	0.0698	0.0698	14.34	0.0699	14.30	1.002	0.9976	1.5010	86°	0'
10'	727	727	13.76	729	13.73	003	974	1.4981		50'
20'	756	756	13.23	758	13.20	003	971	952		40'
30'	0.0785	0.0785	12.75	0.0787	12.71	1.003	0.9969	1.4923		30'
40'	814	814	12.29	816	12.25	003	967	893		20'
50'	844	843	11.87	846	11.83	004	964	864		10'
5° 0'	0.0873	0.0872	11.47	0.0875	11.43	1.004	0.9962	1.4835	85°	0'
10'	902	901	11.10	904	11.06	004	959	806		50'
20'	931	929	10.76	934	10.71	004	957	777		40'
30'	0.0960	0.0958	10.43	0.0963	10.39	1.005	0.9954	1.4748		30'
40'	989	0.0987	10.13	0.0992	10.08	005	951	719		20'
50'	0.1018	0.1016	9.839	0.1022	9.788	005	948	690		10'
6° 0'	0.1047	0.1045	9.567	0.1051	9.514	1.006	0.9945	1.4661	84°	0'
10'	076	074	9.309	080	9.255	006	942	632		50'
20'	105	103	9.065	110	9.010	006	939	603		40'
30'	0.1134	0.1132	8.834	0.1139	8.777	1.006	0.9936	1.4573		30'
40'	164	161	8.614	169	8.556	007	932	544		20'
50'	193	190	8.405	198	8.345	007	929	515		10'
7° 0'	0.1222	0.1219	8.206	0.1228	8.144	1.008	0.9925	1.4486	83°	0'
10'	251	248	8.016	257	7.953	008	922	457		50'
20'	280	276	7.834	287	7.770	008	918	428		40'
30'	0.1309	0.1305	7.661	0.1317	7.596	1.009	0.9914	1.4399		30'
40'	338	334	7.496	346	7.429	009	911	370		20'
50'	367	363	7.337	376	7.269	009	907	341		10'
8° 0'	0.1396	0.1392	7.185	0.1405	7.115	1.010	0.9903	1.4312	82°	0'
10'	425	421	7.040	435	6.968	010	899	283		50'
20'	454	449	6.900	465	6.827	011	894	254		40'
30'	0.1484	0.1478	6.765	0.1495	6.691	1.011	0.9890	1.4224		30'
40'	513	507	6.636	524	6.561	012	886	195		20'
50'	542	536	6.512	554	6.435	012	881	166		10'
9° 0'	0.1571	0.1564	6.392	0.1584	6.314	1.012	0.9877	1.4137	81°	0'
	Cos	Sec	Cot	Tan	Csc	Sin	Radians		Degrees	

233

TABLE A-1 Continued

Degrees	Radians	Sin	Csc	Tan	Cot	Sec	Cos		
9° 0′	0.1571	0.1564	6.392	0.1584	6.314	1.012	0.9877	1.4137	81° 0′
10′	600	593	277	614	197	013	872	108	50′
20′	629	622	166	644	6.084	013	868	079	40′
30′	0.1658	0.1650	6.059	0.1673	5.976	1.014	0.9863	1.4050	30′
40′	687	679	5.955	703	871	014	858	1.4021	20′
50′	716	708	855	733	769	015	853	1.3992	10′
10° 0′	0.1745	0.1736	5.759	0.1763	5.671	1.015	0.9848	1.3963	80° 0′
10′	774	765	665	793	576	016	843	934	50′
20′	804	794	575	823	485	016	838	904	40′
30′	0.1833	0.1822	5.487	0.1853	5.396	1.017	0.9833	1.3875	30′
40′	862	851	403	883	309	018	827	846	20′
50′	891	880	320	914	226	018	822	817	10′
11° 0′	0.1920	0.1908	5.241	0.1944	5.145	1.019	0.9816	1.3788	79° 0′
10′	949	937	164	0.1974	5.066	019	811	759	50′
20′	978	965	089	0.2004	4.989	020	805	730	40′
30′	0.2007	0.1994	5.016	0.2035	4.915	1.020	0.9799	1.3701	30′
40′	036	0.2022	4.945	065	843	021	793	672	20′
50′	065	051	876	095	773	022	787	643	10′
12° 0′	0.2094	0.2079	4.810	0.2126	4.705	1.022	0.9781	1.3614	78° 0′
10′	123	108	745	156	638	023	775	584	50′
20′	153	136	682	186	574	024	769	555	40′
30′	0.2182	0.2164	4.620	0.2217	4.511	1.024	0.9763	1.3526	30′
40′	211	193	560	247	449	025	757	497	20′
50′	240	221	502	278	390	026	750	468	10′
13° 0′	0.2269	0.2250	4.445	0.2309	4.331	1.026	0.9744	1.3439	77° 0′
10′	298	278	390	339	275	027	737	410	50′
20′	327	306	336	370	219	028	730	381	40′
30′	0.2356	0.2334	4.284	0.2401	4.165	1.028	0.9724	1.3352	30′
40′	385	363	232	432	113	029	717	323	20′
50′	414	391	182	462	061	030	710	294	10′
14° 0′	0.2443	0.2419	4.134	0.2493	4.011	1.031	0.9703	1.3265	76° 0′
10′	473	447	086	524	3.962	031	696	235	50′
20′	502	476	4.039	555	914	032	689	206	40′
30′	0.2531	0.2504	3.994	0.2586	3.867	1.033	0.9681	1.3177	30′
40′	560	532	950	617	821	034	674	148	20′
50′	589	560	906	648	776	034	667	119	10′
15° 0′	0.2618	0.2588	3.864	0.2679	3.732	1.035	0.9659	1.3090	75° 0′
10′	647	616	822	711	689	036	652	061	50′
20′	676	644	782	742	647	037	644	032	40′
30′	0.2705	0.2672	3.742	0.2773	3.606	1.038	0.9636	1.3003	30′
40′	734	700	703	805	566	039	628	1.2974	20′
50′	763	728	665	836	526	039	621	945	10′
16° 0′	0.2793	0.2756	3.628	0.2867	3.487	1.040	0.9613	1.2915	74° 0′
10′	822	784	592	899	450	041	605	886	50′
20′	851	812	556	931	412	042	596	857	40′
30′	0.2880	0.2840	3.521	0.2962	3.376	1.043	0.9588	1.2828	30′
40′	909	868	487	0.2994	340	044	580	799	20′
50′	938	896	453	0.3026	305	045	572	770	10′
17° 0′	0.2967	0.2924	3.420	0.3057	3.271	1.046	0.9563	1.2741	73° 0′
10′	996	952	388	089	237	047	555	712	50′
20′	0.3025	0.2979	357	121	204	048	546	683	40′
30′	0.3054	0.3007	3.326	0.3153	3.172	1.048	0.9537	1.2654	30′
40′	083	035	295	185	140	049	528	625	20′
50′	113	062	265	217	108	050	520	595	10′
18° 0′	0.3142	0.3090	3.236	0.3249	3.078	1.051	0.9511	1.2566	72° 0′
		Cos	Sec	Cot	Tan	Csc	Sin	Radians	Degrees

TABLE A-1 Continued

Degrees	Radians	Sin	Csc	Tan	Cot	Sec	Cos		Degrees
18° 0′	0.3142	0.3090	3.236	0.3249	3.078	1.051	0.9511	1.2566	72° 0′
10′	171	118	207	281	047	052	502	537	50′
20′	200	145	179	314	3.018	053	492	508	40′
30′	0.3229	0.3173	3.152	0.3346	2.989	1.054	0.9483	1.2479	30′
40′	258	201	124	378	960	056	474	450	20′
50′	287	228	098	411	932	057	465	421	10′
19° 0′	0.3316	0.3256	3.072	0.3443	2.904	1.058	0.9455	1.2392	71° 0′
10′	345	283	046	476	877	059	446	363	50′
20′	374	311	3.021	508	850	060	436	334	40′
30′	0.3403	0.3338	2.996	0.3541	2.824	1.061	0.9426	1.2305	30′
40′	432	365	971	574	798	062	417	275	20′
50′	462	393	947	607	773	063	407	246	10′
20° 0′	0.3491	0.3420	2.924	0.3640	2.747	1.064	0.9397	1.2217	70° 0′
10′	520	448	901	673	723	065	387	188	50′
20′	549	475	878	706	699	066	377	159	40′
30′	0.3578	0.3502	2.855	0.3739	2.675	1.068	0.9367	1.2130	30′
40′	607	529	833	772	651	069	356	101	20′
50′	636	557	812	805	628	070	346	072	10′
21° 0′	0.3665	0.3584	2.790	0.3839	2.605	1.071	0.9336	1.2043	69° 0′
10′	694	611	769	872	583	072	325	1.2014	50′
20′	723	638	749	906	560	074	315	1.1985	40′
30′	0.3752	0.3665	2.729	0.3939	2.539	1.075	0.9304	1.1956	30′
40′	782	692	709	0.3973	517	076	293	926	20′
50′	811	719	689	0.4006	496	077	283	897	10′
22° 0′	0.3840	0.3746	2.669	0.4040	2.475	1.079	0.9272	1.1868	68° 0′
10′	869	773	650	074	455	080	261	839	50′
20′	898	800	632	108	434	081	250	810	40′
30′	0.3927	0.3827	2.613	0.4142	2.414	1.082	0.9239	1.1781	30′
40′	956	854	595	176	394	084	228	752	20′
50′	985	881	577	210	375	085	216	723	10′
23° 0′	0.4014	0.3907	2.559	0.4245	2.356	1.086	0.9205	1.1694	67° 0′
10′	043	934	542	279	337	088	194	665	50′
20′	072	961	525	314	318	089	182	636	40′
30′	0.4102	0.3987	2.508	0.4348	2.300	1.090	0.9171	1.1606	30′
40′	131	0.4014	491	383	282	092	159	577	20′
50′	160	041	475	417	264	093	147	548	10′
24° 0′	0.4189	0.4067	2.459	0.4452	2.246	1.095	0.9135	1.1519	66° 0′
10′	218	094	443	487	229	096	124	490	50′
20′	247	120	427	522	211	097	112	461	40′
30′	0.4276	0.4147	2.411	0.4557	2.194	1.099	0.9100	1.1432	30′
40′	305	173	396	592	177	100	088	403	20′
50′	334	200	381	628	161	102	075	374	10′
25° 0′	0.4363	0.4226	2.366	0.4603	2.145	1.103	0.9063	1.1345	65° 0′
10′	392	253	352	699	128	105	051	316	50′
20′	422	279	337	734	112	106	038	286	40′
30′	0.4451	0.4305	2.323	0.4770	2.097	1.108	0.9026	1.1257	30′
40′	480	331	309	806	081	109	013	228	20′
50′	509	358	295	841	066	111	0.9001	199	10′
26° 0′	0.4538	0.4384	2.281	0.4877	2.050	1.113	0.8988	1.1170	64° 0′
10′	567	410	268	913	035	114	975	141	50′
20′	596	436	254	950	020	116	962	112	40′
30′	0.4625	0.4462	2.241	0.4986	2.006	1.117	0.8949	1.1083	30′
40′	654	488	228	0.5022	1.991	119	936	054	20′
50′	683	514	215	059	977	121	923	1.1025	10′
27° 0′	0.4712	0.4540	2.203	0.5095	1.963	1.122	0.8910	1.0996	63° 0′
		Cos	Sec	Cot	Tan	Csc	Sin	Radians	Degrees

TABLE A-1 Continued

Degrees	Radians	Sin	Csc	Tan	Cot	Sec	Cos	Radians	Degrees
27° 0'	0.4712	0.4540	2.203	0.5095	1.963	1.122	0.8910	1.0996	63° 0'
10'	741	566	190	132	949	124	897	966	50'
20'	771	592	178	169	935	126	884	937	40'
30'	0.4800	0.4617	2.166	0.5206	1.921	1.127	0.8870	1.0908	30'
40'	829	643	154	243	907	129	857	879	20'
50'	858	669	142	280	894	131	843	850	10'
28° 0'	0.4887	0.4695	2.130	0.5317	1.881	1.133	0.8829	1.0821	62° 0'
10'	916	720	118	354	868	134	816	792	50'
20'	945	746	107	392	855	136	802	763	40'
30'	0.4974	0.4772	2.096	0.5430	1.842	1.138	0.8788	1.0734	30'
40'	0.5003	797	0.85	467	829	140	774	705	20'
50'	032	823	074	505	816	142	760	676	10'
29° 0'	0.5061	0.4848	2.063	0.5543	1.804	1.143	0.8746	1.0647	61° 0'
10'	091	874	052	581	792	145	732	617	50'
20'	120	899	041	691	780	147	718	588	40'
30'	0.5149	0.4924	2.031	0.5658	1.767	1.149	0.8704	1.0559	30'
40'	178	950	020	696	756	151	689	530	20'
50'	207	0.4975	010	735	744	153	675	501	10'
30° 0'	0.5236	0.5000	2.000	0.5774	1.732	1.155	0.8660	1.0472	60° 0'
10'	265	025	1.990	812	720	157	646	443	50'
20'	294	050	980	851	709	159	631	414	40'
30'	0.5323	0.5075	1.970	0.5890	1.698	1.161	0.8616	1.0385	30'
40'	352	100	961	930	686	163	601	356	20'
50'	381	125	951	0.5969	675	165	587	327	10'
31° 0'	0.5411	0.5150	1.942	0.6009	1.664	1.167	0.8572	1.0297	59° 0'
10'	440	175	932	048	653	169	557	268	50'
20'	469	200	923	088	643	171	542	239	40'
30'	0.5498	0.5225	1.914	0.6128	1.632	1.173	0.8526	1.0210	30'
40'	527	250	905	168	621	175	511	181	20'
50'	556	275	896	208	611	177	496	152	10'
32° 0'	0.5585	0.5299	1.887	0.6249	1.600	1.179	0.8480	1.0123	58° 0'
10'	614	324	878	289	590	181	465	094	50'
20'	643	348	870	330	580	184	450	065	40'
30'	0.5672	0.5373	1.861	0.6371	1.570	1.186	0.8434	1.0036	30'
40'	701	398	853	412	560	188	418	1.0007	20'
50'	730	422	844	453	550	190	403	0.9977	10'
33° 0'	0.5760	0.5446	1.836	0.6494	1.540	1.192	0.8387	0.9948	57° 0'
10'	789	471	828	536	530	195	371	919	50'
20'	818	495	820	577	520	197	355	890	40'
30'	0.5847	0.5519	1.812	0.6619	1.511	1.199	0.8339	0.9861	30'
40'	876	544	804	661	501	202	323	832	20'
50'	905	568	796	703	1.492	204	307	803	10'
34° 0'	0.5934	0.5592	1.788	0.6745	1.483	1.206	0.8290	0.9774	56° 0'
10'	963	616	781	787	473	209	274	745	50'
20'	992	640	773	830	464	211	258	716	40'
30'	0.6021	0.5664	1.766	0.6873	1.455	1.213	0.8241	0.9687	30'
40'	050	688	758	916	446	216	225	657	20'
50'	080	712	751	0.6959	437	218	208	628	10'
35° 0'	0.6109	0.5736	1.743	0.7002	1.428	1.221	0.8192	0.9599	55° 0'
10'	138	760	736	046	419	223	175	570	50'
20'	167	783	729	089	411	226	158	541	40'
30'	0.6196	0.5807	1.722	0.7133	1.402	1.228	0.8141	0.9512	30'
40'	225	831	715	177	393	231	124	483	20'
50'	254	854	708	221	385	233	107	454	10'
36° 0'	0.6283	0.5878	1.701	0.7265	1.376	1.236	0.8090	0.9425	54° 0'
		Cos	Sec	Cot	Tan	Csc	Sin	Radians	Degrees

TABLE A-1 Continued

Degrees	Radians	Sin	Csc	Tan	Cot	Sec	Cos			
36° 0'	0.6283	0.5878	1.701	0.7265	1.376	1.236	0.8090	0.9425	54°	0'
10'	312	901	695	310	368	239	073	396		50'
20'	341	925	688	355	360	241	056	367		40'
30'	0.6370	0.5948	1.681	0.7400	1.351	1.244	0.8039	0.9338		30'
40'	400	972	675	445	343	247	021	308		20'
50'	429	0.5995	668	490	335	249	0.8004	279		10'
37° 0'	0.6458	0.6018	1.662	0.7536	1.327	1.252	0.7986	0.9250	53°	0'
10'	487	041	655	581	319	255	969	221		50'
20'	516	065	649	627	311	258	951	192		40'
30'	0.6545	0.6088	1.643	0.7673	1.303	1.260	0.7934	0.9163		30'
40'	574	111	636	720	295	263	916	134		20'
50'	603	134	630	766	288	266	898	105		10'
38° 0'	0.6632	0.6157	1.624	0.7813	1.280	1.269	0.7880	0.9076	52°	0'
10'	661	180	618	860	272	272	862	047		50'
20'	690	202	612	907	265	275	844	0.9018		40'
30'	0.6720	0.6225	1.606	0.7954	1.257	1.278	0.7826	0.8988		30'
40'	749	248	601	0.8002	250	281	808	959		20'
50'	778	271	595	050	242	284	790	930		10'
39° 0'	0.6807	0.6293	1.589	0.8098	1.235	1.287	0.7771	0.8901	51°	0'
10'	836	316	583	146	228	290	753	872		50'
20'	865	338	578	195	220	293	735	843		40'
30'	0.6894	0.6361	1.572	0.8243	1.213	1.296	0.7716	0.8814		30'
40'	923	383	567	292	206	299	698	785		20'
50'	952	406	561	342	199	302	679	756		10'
40° 0'	0.6981	0.6428	1.556	0.8391	1.192	1.305	0.7660	0.8727	50°	0'
10'	0.7010	450	550	441	185	309	642	698		50'
20'	039	472	545	491	178	312	623	668		40'
30'	0.7069	0.6494	1.540	0.8541	1.171	1.315	0.7604	0.8639		30'
40'	098	517	535	591	164	318	585	610		20'
50'	127	539	529	642	157	322	566	581		10'
41° 0'	0.7156	0.6561	1.524	0.8693	1.150	1.325	0.7547	0.8552	49°	0'
10'	185	583	519	744	144	328	528	523		50'
20'	214	604	514	796	137	332	509	494		40'
30'	0.7243	0.6626	1.509	0.8847	1.130	1.335	0.7490	0.8465		30'
40'	272	648	504	899	124	339	470	436		20'
50'	301	670	499	0.8952	117	342	451	407		10'
42° 0'	0.7330	0.6691	1.494	0.9004	1.111	1.346	0.7431	0.8378	48°	0'
10'	359	713	490	057	104	349	412	348		50'
20'	389	734	485	110	098	353	392	319		40'
30'	0.7418	0.6756	1.480	0.9163	1.091	1.356	0.7373	0.8290		30'
40'	447	777	476	217	085	360	353	261		20'
50'	476	799	471	271	079	364	333	232		10'
43° 0'	0.7505	0.6820	1.466	0.9325	1.072	1.367	0.7314	0.8203	47°	0'
10'	534	841	462	380	066	371	294	174		50'
20'	563	862	457	435	060	375	274	145		40'
30'	0.7592	0.6884	1.453	0.9490	1.054	1.379	0.7254	0.8116		30'
40'	621	905	448	545	048	382	234	087		20'
50'	650	926	444	601	042	386	214	058		10'
44° 0'	0.7679	0.6947	1.440	0.9657	1.036	1.390	0.7193	0.8029	46°	0'
10'	709	967	435	713	030	394	173	0.7999		50'
20'	738	0.6988	431	770	024	398	153	970		40'
30'	0.7767	0.7009	1.427	0.9827	1.018	1.402	0.7133	0.7941		30'
40'	796	030	423	884	012	406	112	912		20'
50'	825	050	418	0.9942	006	410	092	883		10'
45° 0'	0.7854	0.7071	1.414	1.000	1.000	1.414	0.7071	0.7854	45°	0'
		Cos	Sec	Cot	Tan	Csc	Sin	Radians	Degrees	

TABLE A-2: FUNCTIONS FOR ANGLES GREATER THAN 90°

Angle	Sine	Cosine	Tangent	Cotangent
$\emptyset$	$+\sin \emptyset$	$+\cos \emptyset$	$+\tan \emptyset$	$+\cot \emptyset$
$90° + \emptyset$	$+\cos \emptyset$	$-\sin \emptyset$	$-\cot \emptyset$	$-\tan \emptyset$
$180° + \emptyset$	$-\sin \emptyset$	$-\cos \emptyset$	$+\tan \emptyset$	$+\cot \emptyset$
$270° + \emptyset$	$-\cos \emptyset$	$+\sin \emptyset$	$-\cot \emptyset$	$-\tan \emptyset$

Example: sin of 132°

$132° - 90° = 42°$

$\sin 132° = +\cos 42° = 0.74314$

TABLE A-3: POWERS, ROOTS, AND RECIPROCALS 1 THROUGH 100

n	n^2	n^3	$\sqrt{n}$	$\sqrt[3]{n}$	$1/n$	n	n^2	n^3	$\sqrt{n}$	$\sqrt[3]{n}$	$1/n$
1	1	1	1.000	1.000	1.0000	51	2,601	132,651	7.141	3.708	0.0196
2	4	8	1.414	1.260	0.5000	52	2,704	140,608	7.211	3.733	0.0192
3	9	27	1.732	1.442	0.3333	53	2,809	148,877	7.280	3.756	0.0189
4	16	64	2.000	1.587	0.2500	54	2,916	157,464	7.348	3.780	0.0185
5	25	125	2.236	1.710	0.2000	55	3,025	166,375	7.416	3.803	0.0182
6	36	216	2.449	1.817	0.1667	56	3,136	175,616	7.483	3.826	0.0179
7	49	343	2.646	1.913	0.1429	57	3,249	185,193	7.550	3.849	0.0175
8	64	512	2.828	2.000	0.1250	58	3,364	195,112	7.616	3.871	0.0172
9	81	729	3.000	2.080	0.1111	59	3,481	205,379	7.681	3.893	0.0169
10	100	1,000	3.162	2.154	0.1000	60	3,600	216,000	7.746	3.915	0.0167
11	121	1,331	3.317	2.224	0.0909	61	3,721	226,981	7.810	3.936	0.0164
12	144	1,728	3.464	2.289	0.0833	62	3,844	238,328	7.874	3.958	0.0161
13	169	2,197	3.606	2.351	0.0769	63	3,969	250,047	7.937	3.979	0.0159
14	196	2,744	3.742	2.410	0.0714	64	4,096	262,144	8.000	4.000	0.0156
15	225	3,375	3.873	2.466	0.0667	65	4,225	274,625	8.062	4.021	0.0154
16	256	4,096	4.000	2.520	0.0625	66	4,356	287,496	8.124	4.041	0.0152
17	289	4,913	4.123	2.571	0.0588	67	4,489	300,763	8.185	4.062	0.0149
18	324	5,832	4.243	2.621	0.0556	68	4,624	314,432	8.246	4.082	0.0147
19	361	6,859	4.359	2.668	0.0526	69	4,761	328,509	8.307	4.102	0.0145
20	400	8,000	4.472	2.714	0.0500	70	4,900	343,000	8.367	4.121	0.0143
21	441	9,261	4.583	2.759	0.0476	71	5,041	357,911	8.426	4.141	0.0141
22	484	10,648	4.690	2.802	0.0455	72	5,184	373,248	8.485	4.160	0.0139
23	529	12,167	4.796	2.844	0.0435	73	5,329	389,017	8.544	4.179	0.0137
24	576	13,824	4.899	2.884	0.0417	74	5,476	405,224	8.602	4.198	0.0135
25	625	15,625	5.000	2.924	0.0400	75	5,625	421,875	8.660	4.217	0.0133
26	676	17,576	5.099	2.962	0.0385	76	5,776	438,976	8.718	4.236	0.0132
27	729	19,683	5.196	3.000	0.0370	77	5,929	456,533	8.775	4.254	0.0130
28	784	21,952	5.292	3.037	0.0357	78	6,084	474,552	8.832	4.273	0.0128
29	841	24,389	5.385	3.072	0.0345	79	6,241	493,039	8.888	4.291	0.0127
30	900	27,000	5.477	3.107	0.0333	80	6,400	512,000	8.944	4.309	0.0125
31	961	29,791	5.568	3.141	0.0323	81	6,561	531,441	9.000	4.327	0.0123
32	1,024	32,768	5.657	3.175	0.0312	82	6,724	551,368	9.055	4.344	0.0122
33	1,089	35,937	5.745	3.208	0.0303	83	6,889	571,787	9.110	4.362	0.0120
34	1,156	39,304	5.831	3.240	0.0294	84	7,056	592,704	9.165	4.380	0.0119
35	1,225	42,875	5.916	3.271	0.0286	85	7,225	614,125	9.220	4.397	0.0118
36	1,296	46,656	6.000	3.302	0.0278	86	7,396	636,056	9.274	4.414	0.0116
37	1,369	50,653	6.083	3.332	0.0270	87	7,569	658,503	9.327	4.431	0.0115
38	1,444	54,872	6.164	3.362	0.0263	88	7,744	681,472	9.381	4.448	0.0114
39	1,521	59,319	6.245	3.391	0.0256	89	7,921	704,969	9.434	4.465	0.0112
40	1,600	64,000	6.325	3.420	0.0250	90	8,100	729,000	9.487	4.481	0.0111
41	1,681	68,921	6.403	3.448	0.0244	91	8,281	753,571	9.539	4.498	0.0110
42	1,764	74,088	6.481	3.476	0.0238	92	8,464	778,688	9.592	4.514	0.0109
43	1,849	79,507	6.557	3.503	0.0233	93	8,649	804,357	9.644	4.531	0.0108
44	1,936	85,184	6.633	3.530	0.0227	94	8,836	830,584	9.695	4.547	0.0106
45	2,025	91,125	6.708	3.557	0.0222	95	9,025	857,375	9.747	4.563	0.0105
46	2,116	97,336	6.782	3.583	0.0217	96	9,216	884,736	9.798	4.579	0.0104
47	2,209	103,823	6.856	3.609	0.0213	97	9,409	912,673	9.849	4.595	0.0103
48	2,304	110,592	6.928	3.634	0.0208	98	9,604	941,192	9.899	4.610	0.0102
49	2,401	117,649	7.000	3.659	0.0204	99	9,801	970,299	9.950	4.626	0.0101
50	2,500	125,000	7.071	3.684	0.0200	100	10,000	1,000,000	10.000	4.642	0.0100

TABLE A-4: CONSTANTS

Constant	Value
Pi (π), the ratio of circumference to diameter	3.14159
Base of hyperbolic logarithms	2.71828
Modulus of common system of logs	0.43429
Acceleration due to gravity at NY	32.15949
Pounds per square inch (psi) due to 1 atmosphere	14.7
Pounds per square inch due to 1 foot head of water	0.434
Feet of head for pressure of 1 pound per square inch	2.304

Approximate Values of Sines

Natural sine of 1°	$\dfrac{1.75 \text{ foot}}{100 \text{ feet}} = 0.0175 = \dfrac{1}{60} \text{ (approx.)}$
Natural sine of 0° 1'	$\dfrac{0.03 \text{ foot}}{100 \text{ feet}}$
Natural sine of 0° 00' 01"	$\dfrac{0.3 \text{ inch}}{1 \text{ mile}}$

TABLE A-5: LENGTHS OF CIRCULAR ARCS. RADIUS = 1

Sec.	Length.	Min.	Length.	Deg.	Length.	Deg.	Length.
1	0.0000048	1	0.0002909	1	0.0174533	61	1.0646508
2	0.0000097	2	0.0005818	2	0.0349066	62	1.0821041
3	0.0000145	3	0.0008727	3	0.0523599	63	1.0995574
4	0.0000194	4	0.0011636	4	0.0698132	64	1.1170107
5	0.0000242	5	0.0014544	5	0.0872665	65	1.1344640
6	0.0000291	6	0.0017453	6	0.1047198	66	1.1519173
7	0.0000339	7	0.0020362	7	0.1221730	67	1.1693706
8	0.0000388	8	0.0023271	8	0.1396263	68	1.1868239
9	0.0000436	9	0.0026180	9	0.1570796	69	1.2042772
10	0.0000485	10	0.0029089	10	0.1745329	70	1.2217305
11	0.0000533	11	0.0031998	11	0.1919862	71	1.2391838
12	0.0000582	12	0.0034907	12	0.2094395	72	1.2566371
13	0.0000630	13	0.0037815	13	0.2268928	73	1.2740904
14	0.0000679	14	0.0040724	14	0.2443461	74	1.2915436
15	0.0000727	15	0.0043633	15	0.2617994	75	1.3089969
16	0.0000776	16	0.0046542	16	0.2792527	76	1.3264502
17	0.0000824	17	0.0049451	17	0.2967060	77	1.3439035
18	0.0000873	18	0.0052360	18	0.3141593	78	1.3613568
19	0.0000921	19	0.0055269	19	0.3316126	79	1.3788101
20	0.0000970	20	0.0058178	20	0.3490659	80	1.3962634
21	0.0001018	21	0.0061087	21	0.3665191	81	1.4137167
22	0.0001067	22	0.0063995	22	0.3839724	82	1.4311700
23	0.0001115	23	0.0066904	23	0.4014257	83	1.4486233
24	0.0001164	24	0.0069813	24	0.4188790	84	1.4660766
25	0.0001212	25	0.0072722	25	0.4363323	85	1.4835299
26	0.0001261	26	0.0075631	26	0.4537856	86	1.5009832
27	0.0001309	27	0.0078540	27	0.4712389	87	1.5184364
28	0.0001357	28	0.0081449	28	0.4886922	88	1.5358897
29	0.0001406	29	0.0084358	29	0.5061455	89	1.5533430
30	0.0001454	30	0.0087266	30	0.5235988	90	1.5707963
31	0.0001503	31	0.0090175	31	0.5410521	91	1.5882496
32	0.0001551	32	0.0093084	32	0.5585054	92	1.6057029
33	0.0001600	33	0.0095993	33	0.5759587	93	1.6231562
34	0.0001648	34	0.0098902	34	0.5934119	94	1.6406095
35	0.0001697	35	0.0101811	35	0.6108652	95	1.6580628
36	0.0001745	36	0.0104720	36	0.6283185	96	1.6755161
37	0.0001794	37	0.0107629	37	0.6457718	97	1.6929694
38	0.0001842	38	0.0110538	38	0.6632251	98	1.7104227
39	0.0001891	39	0.0113446	39	0.6806784	99	1.7278760
40	0.0001939	40	0.0116355	40	0.6981317	100	1.7453293
41	0.0001988	41	0.0119264	41	0.7155850	101	1.7627825
42	0.0002036	42	0.0122173	42	0.7330383	102	1.7802358
43	0.0002085	43	0.0125082	43	0.7504916	103	1.7976891
44	0.0002133	44	0.0127991	44	0.7679449	104	1.8151424
45	0.0002182	45	0.0130900	45	0.7853982	105	1.8325957
46	0.0002230	46	0.0133809	46	0.8028515	106	1.8500490
47	0.0002279	47	0.0136717	47	0.8203047	107	1.8675023
48	0.0002327	48	0.0139626	48	0.8377580	108	1.8849556
49	0.0002376	49	0.0142535	49	0.8552113	109	1.9024089
50	0.0002424	50	0.0145444	50	0.8726646	110	1.9198622
51	0.0002473	51	0.0148353	51	0.8901179	111	1.9373155
52	0.0002521	52	0.0151262	52	0.9075712	112	1.9547688
53	0.0002570	53	0.0154171	53	0.9250245	113	1.9722221
54	0.0002618	54	0.0157080	54	0.9424778	114	1.9896753
55	0.0002666	55	0.0159991	55	0.9599311	115	2.0071286
56	0.0002715	56	0.0162897	56	0.9773844	116	2.0245819
57	0.0002763	57	0.0165806	57	0.9948377	117	2.0420352
58	0.0002812	58	0.0168715	58	1.0122910	118	2.0594885
59	0.0002860	59	0.0171624	59	1.0297443	119	2.0769418
60	0.0002909	60	0.0174533	60	1.0471976	120	2.0943951

TABLE A-6: NATURAL VERSED SINES

			3°	4°	5°	6°	7°	8°	9°
0° 00'	0.00000	00'	0.00137	0.00244	0.00381	0.00548	0.00745	0.00973	0.01231
05	000	01	139	246	383	551	749	977	236
10	000	02	140	248	386	554	752	981	240
15	001	03	142	250	388	557	756	985	245
20	002	04	143	252	391	560	760	989	249
25	003	05	0.00145	0.00254	0.00393	0.00563	0.00763	0.00994	0.01254
30	0.00004	06	146	256	396	566	767	998	259
35	005	07	148	258	398	569	770	0.01002	263
40	007	08	149	260	401	572	774	006	268
45	009	09	151	262	404	576	778	010	272
50	011	10	0.00153	0.00264	0.00406	0.00579	0.00781	0.01014	0.01277
55	013	11	154	266	409	582	785	018	282
1° 00'	0.00015	12	156	269	412	585	789	022	286
05	018	13	158	271	414	588	792	027	291
10	021	14	159	273	417	591	796	031	296
15	024	15	0.00161	0.00275	0.00420	0.00594	0.00800	0.01035	0.01300
20	027	16	162	277	422	598	803	039	305
25	031	17	164	279	425	601	807	043	310
30	0.00034	18	166	281	423	604	811	047	314
35	038	19	167	284	430	607	814	052	319
40	042	20	0.00169	0.00286	0.00433	0.00610	0.00818	0.01056	0.01324
45	047	21	171	288	436	614	822	060	329
50	051	22	173	290	438	617	825	064	333
55	056	23	174	292	441	620	829	069	338
2° 00'	0.00061	24	176	295	444	623	833	073	343
		25	0.00178	0.00297	0.00447	0.00626	0.00837	0.01077	0.01348
		26	179	299	449	630	840	081	352
		27	181	301	452	633	844	086	357
		28	183	304	455	636	848	090	362
		29	185	306	458	640	852	094	367
2° 00'	0.00061	30	0.00187	0.00308	0.00460	0.00643	0.00856	0.01098	0.01371
02	063	31	188	311	463	646	859	103	376
04	065	32	190	313	466	649	863	107	381
06	067	33	192	315	469	653	867	111	386
08	069	34	194	317	472	656	871	116	391
10	0.00071	35	0.00196	0.00320	0.00474	0.00659	0.00875	0.01120	0.01396
12	074	36	197	322	477	663	878	124	400
14	076	37	199	324	480	666	882	129	405
16	078	38	201	327	483	669	886	133	410
18	081	39	203	329	486	673	890	137	415
20	0.00083	40	0.00205	0.00332	0.00489	0.00676	0.00894	0.01142	0.01420
22	085	41	207	334	492	680	898	146	425
24	088	42	208	336	494	683	902	151	430
26	090	43	210	339	497	686	906	155	435
28	093	44	212	341	500	690	909	159	439
30	0.00095	45	0.00214	0.00343	0.00503	0.00693	0.00913	0.01164	0.01444
32	098	46	216	346	506	697	917	168	449
34	100	47	218	348	509	700	921	173	454
36	103	48	220	351	512	703	925	177	459
38	106	49	222	353	515	707	929	182	464
40	0.00108	50	0.00224	0.00356	0.00518	0.00710	0.00933	0.01186	0.01469
42	111	51	226	358	521	714	937	191	474
44	114	52	228	361	524	717	941	195	479
46	117	53	230	363	527	721	945	200	484
48	120	54	232	365	530	724	949	204	489
50	0.00122	55	0.00234	0.00368	0.00533	0.00728	0.00953	0.01209	0.01494
52	125	56	236	370	536	731	957	213	499
54	128	57	238	373	539	735	961	218	504
56	131	58	240	375	542	738	965	222	509
58	134	59	0242	378	545	742	969	227	514
3° 00'	0.00137	60	0.0244	0.00381	0.00548	0.00745	0.00973	0.01231	0.01519

	10°	11°	12°	13°	14°	15°	16°	17°
00'	0.01519	0.01837	0.02185	0.02563	0.02970	0.03407	0.03874	0.04370
01	524	843	191	570	977	415	882	378
02	529	848	197	576	985	422	890	387
03	534	854	203	583	992	430	898	395
04	539	860	209	589	999	438	906	404
05	0.01545	0.01865	0.02216	0.02596	0.03006	0.03445	0.03914	0.04412
06	550	871	222	602	013	453	922	421
07	555	876	228	609	020	460	930	429
08	560	882	234	616	027	468	938	438
09	565	888	240	622	034	476	946	446
10	0.01570	0.01893	0.02246	0.02629	0.03041	0.03483	0.03954	0.04455
11	575	899	252	635	048	491	963	464
12	580	904	258	642	055	498	971	472
13	586	910	265	649	063	506	979	481
14	591	916	271	655	070	514	987	489
15	0.01596	0.01921	0.02277	0.02662	0.03077	0.03521	0.03995	0.04498
16	601	927	283	669	084	529	0.04003	507
17	606	933	289	675	091	537	011	515
18	611	939	295	682	098	544	019	524
19	617	944	302	689	106	552	028	533
20	0.01622	0.01950	0.02308	0.02696	0.03113	0.03560	0.04036	0.04541
21	627	956	314	702	120	567	044	550
22	632	961	320	709	127	575	052	559
23	638	967	327	716	134	583	060	567
24	643	973	333	722	142	590	069	576
25	0.01648	0.01979	0.02339	0.02729	0.03149	0.03598	0.04077	0.04585
26	653	984	345	736	156	606	085	593
27	659	990	352	743	163	614	093	602
28	664	996	358	749	171	621	102	611
29	669	0.02002	364	756	178	629	110	620
30	0.01675	0.02008	0.02370	0.02763	0.03185	0.03637	0.04118	0.04628
31	680	013	377	770	193	645	126	637
32	685	019	383	777	200	653	135	646
33	690	025	389	783	207	660	143	655
34	696	031	396	790	214	668	151	663
35	0.01701	0.02037	0.02402	0.02797	0.03222	0.03676	0.04159	0.04672
36	706	042	408	804	229	684	168	681
37	712	048	415	811	236	692	176	690
38	717	054	421	818	244	699	184	699
39	723	060	427	824	251	707	193	707
40	0.01728	0.02066	0.02434	0.02831	0.03258	0.03715	0.04201	0.04716
41	733	072	440	838	266	723	209	725
42	739	078	447	845	273	731	218	734
43	744	084	453	852	281	739	226	743
44	750	090	459	859	288	747	234	752
45	0.01755	0.02095	0.02466	0.02866	0.03295	0.03754	0.04243	0.04760
46	760	101	472	873	303	762	251	769
47	766	107	479	880	310	770	260	778
48	771	113	485	887	318	778	268	787
49	777	119	492	894	325	786	276	796
50	0.01782	0.02125	0.02498	0.02900	0.03333	0.03794	0.04285	0.04805
51	788	131	504	907	340	802	293	814
52	793	137	511	914	347	810	302	823
53	799	143	517	921	355	818	310	832
54	804	149	524	928	362	826	319	841
55	0.01810	0.02155	0.02530	0.02935	0.03370	0.03834	0.04327	0.04850
56	815	161	537	942	377	842	336	858
57	821	167	543	949	385	850	344	867
58	826	173	550	956	392	858	353	876
59	832	179	556	963	400	866	361	885
60	0.01837	0.02185	0.02563	0.02970	0.03407	0.03874	0.04370	0.04894

A variety of formulae can be extremely useful to map and topographic drafters. These formulae are broken down into three major groupings: 1) computation of regular areas, volumes, surfaces; 2) computation of irregular areas and volumes; and 3) triangle, arc, and chord computations.

COMPUTATION OF AREAS AND VOLUMES FOR REGULAR SURFACES

Area and volume computations are useful in figuring land area and amount of cut and fill. These include the geometric properties of plane figures, volumes and surfaces of double-curved solids, and volumes and surfaces of typical solids.

Geometric Properties of Plane Figures
1. Triangle (Figure A-1):

$$\text{Area} = 0.5 \, (\text{altitude}) \times \text{base} = \frac{b \times h}{2}$$

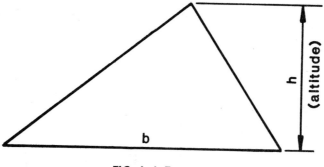

FIG. A-1 Triangle

2. Trapezoid (Figure A-2):

$$Area = 0.5 (\text{sum of parallel sides} \times \text{altitude} = \frac{h(a + b)}{2}$$

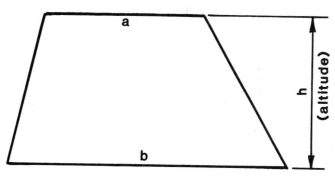

FIG. A-2 Trapezoid

3. Trapezium or Irregular Quadrilateral (Figure A-3):

Area = divide the figure into 2 triangles and sum their area

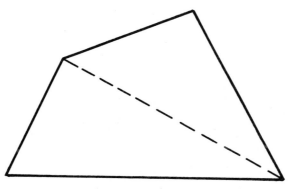

FIG. A-3 Trapezium

4. Parallelogram (Figure A-4):

$$\text{Area} = \text{side} \times \text{altitude} = b \times h$$

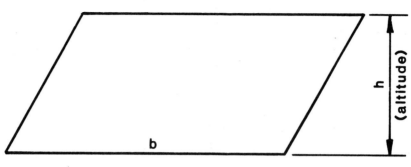

FIG. A-4 Parallelogram

5. Circle (Figure A-5):

$$\text{Area} = \pi R^2 = \frac{\pi D^2}{4}$$

where: $\pi = 3.1415$

$$\text{Circumference} = 2\pi R = \pi D$$

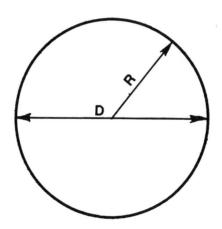

FIG. A-5 Circle

6. Regular Polygon (Figure A-6):

$$\text{Area} = \frac{nSR_1}{2}$$

where: n = number of sides

$$R_1 = \frac{S}{2 \tan \phi}$$

$$R_2 = \frac{S}{2 \sin \phi}$$

Any side $S = 2\sqrt{R_1^2 - R_2^2}$

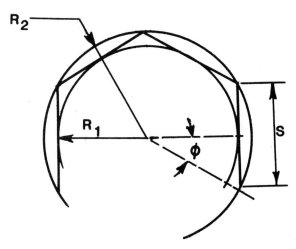

FIG. A-6 Regular polygon

7. Circular Segment (Figure A-7):

$$\text{Area} = \frac{(\text{length of arc } a \times R - c(R-y))}{2}$$

$$\text{Chord } c = 2\sqrt{2yR - y^2} = \frac{2R \sin A°}{2}$$

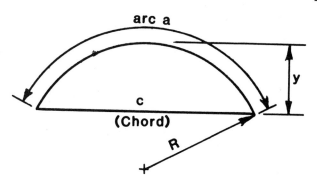

Fig. A-7 Circular segment

8. Circular Sector (Figure A-8):

$$\text{Area} = 0.5(R \times \text{length of arc } a)$$

$$= \frac{\text{area of the circle} \times A°}{360°}$$

$$= 0.0087R^2A°$$

$$\text{Arc } a = 0.0175RA° = \frac{\pi RA°}{180°}$$

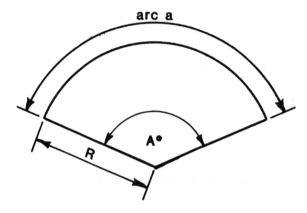

FIG. A-8 Circular sector

Volumes and Surfaces of Double-curved Solids

1. Sphere (Figure A-9):

$$\text{Volume} = 0.5236D^3 = \frac{4\pi R^3}{3}$$

$$\text{Surface} = 4\pi R^2 = \pi D^2$$

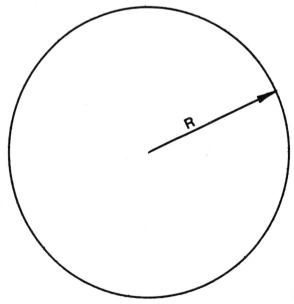

FIG. A-9 Sphere

2. Ellipsoid (Figure A-10):

$$\text{Volume} = \frac{\pi abc}{6}$$

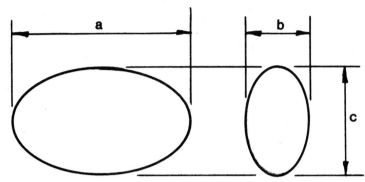

FIG. A-10 Ellipsoid

Volumes and Surfaces of Double-curved Solids
1. Prisms and Cylinders (Figure A-11):

Volume = altitude × area of base

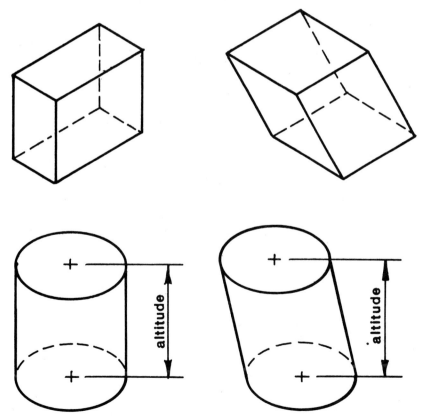

FIG. A-11 Prisms and cylinders

2. Pyramids and Cones (Figure A-12):

$$\text{Volume} = \frac{\text{area of base} \times \text{altitude}}{3}$$

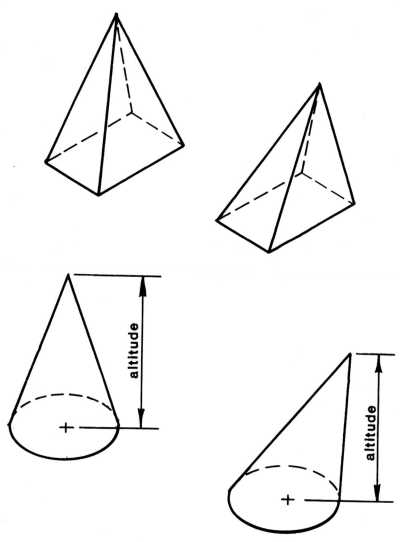

FIG. A-12 Pyramids and cones

COMPUTATION OF IRREGULAR AREAS AND VOLUMES

Since most topographic drawings deal with land areas that are
irregularly shaped, the computations presented in this section would

prove to be extremely valuable. Included in this section are three types of computations: the area of an irregular plane figure, the volume of an irregular figure, and the volume of cut and fill from a contour drawing.

Area of an Irregular Plane Figure. To calculate the area of an irregular plane (Figure A-13), divide the figure into parallel sections by parallel lines that are equally spaced, and measure the length of each parallel line. Obtain the area by using either the *Trapezoid Rule* or *Durand's Rule:*

- **Trapezoid Rule.** Sum the length of each parallel line, except for the first and last parallel line, which are taken at half their length, and multiply the total sum by the distance between the parallel lines.
  Note: Use this rule for estimating most irregular plane areas.
- **Durand's Rule.** Find the sum of each parallel line at the following values: the first and last line at 5/12 value, the next lines in at 13/12 value, and all other parallel lines at true value. Multiply the sum total by the distance between the parallel lines.
  Note: Use this rule for irregular shapes that are exaggerated.

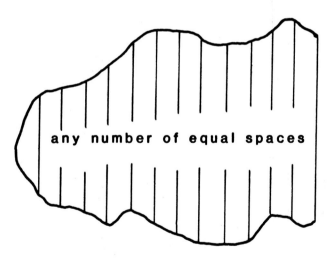

any number of equal spaces

FIG. A-13 Calculating area of irregular plane figure

Volume of an Irregular Figure. The method used for computing the volume of an irregular figure is sometimes referred to as the *sectioning method,* because the figure is divided into equal sections. The procedure is as follows (see Figure A-14):

251

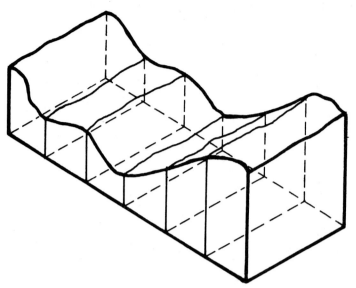

FIG. A-14 Calculating volume of irregular figure

1. Divide the figure into a series of equally spaced sections.

2. Calculate the area of each section by one of the methods previously discussed.

3. Calculate the volume using either the Trapezoid Rule or Durand's Rule. This will give the total volume.

Volume of Cut and Fill from a Contour Drawing. Since *cut and fill* refers to the amount of land to be excavated and filled, the contour drawing should show both the original and finished contours of the area of land (Figure A-15). The procedure for calculating the volume of cut and fill is as follows:

1. Determine the difference in areas between the new and old contours, and enter the values in a chart for cut and fill.

2. Total each column to determine total volume of cut and fill. *Note:* When cut or fill ends on a contour, use half value.

TRIANGLE, ARC, AND CHORD COMPUTATIONS

The procedures used to make triangle, arc, and chord computations are based upon basic trigonometric principles. Computations are divided into four areas: right triangles, oblique triangles, arcs, and chords.

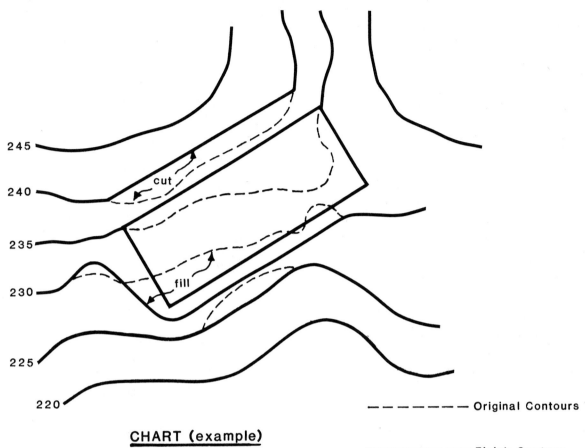

245

240   cut

235

230   fill

225

220

– – – – – – – Original Contours

―――――― Finish Contour

CHART (example)

CONTOUR	FILL	CUT
245		312
240		965
235	3,800 ÷ 2 = 1900	2,460 ÷ 2 = 1230
230	2,265	20
	7,115 × 5	10,010 × 5
TOTALS:	35,575 cu. ft.	50,050 cu. ft

FIG. A-15 Example of cut and fill problem

Right Triangles. These are any triangles that have one angle equal to 90°. The basic formulae used in right triangle calculations are given here (Figure A-16).

$$A + B + C = 180° \qquad\qquad A + B = 90°$$

$$\tan A = \frac{a}{b} \qquad\qquad \sin A = \frac{a}{c} \qquad\qquad \cos A = \frac{b}{c}$$

$$\cot A = \frac{b}{a} \qquad\qquad \csc A = \frac{c}{a} \qquad\qquad \sec A = \frac{c}{b}$$

$$\text{Versed } \sin A = \frac{(c - b)}{c} = 1 - \cos A$$

$$\tan A = \cot B \qquad\qquad \sin A = \cos B \qquad\qquad \cos A = \sin B$$

$$\cot A = \tan B \qquad\qquad \csc A = \sec B \qquad\qquad \sec A = \csc B$$

$$a^2 = c^2 - b^2 \qquad\qquad b^2 = c^2 - a^2 \qquad\qquad c^2 = a^2 + b^2$$

$$a = \sqrt{(c + b)(c - b)} \qquad b = \sqrt{(c + a)(c - a)} \qquad c = \sqrt{a^2 + b^2}$$

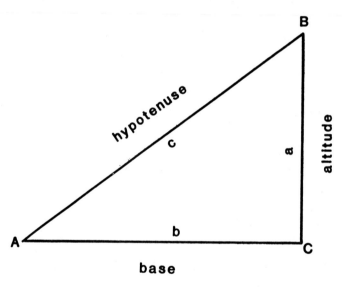

FIG. A-16 Right triangle

Oblique Triangles. These are any triangles that have no angle equal to 90°. The basic formulae used in oblique triangle calculations are given here (Figure A-17).

$$A + B + C = 180°$$

$$\text{Law of Sines:} \quad \frac{a}{\sin A} = \frac{b}{\sin B} = \frac{c}{\sin C}$$

$$a = \frac{b\,(\sin A)}{\sin B} = \frac{c\,(\sin A)}{\sin C}$$

$$b = \frac{a\,(\sin B)}{\sin A} = \frac{c\,(\sin B)}{\sin C}$$

$$c = \frac{a\,(\sin C)}{\sin A} = \frac{b\,(\sin C)}{\sin B}$$

Law of Cosines: $a^2 = b^2 + c^2 - 2bc\,(\cos A)$

$$b^2 = a^2 + c^2 - 2ac\,(\cos B)$$

$$c^2 = a^2 + b^2 - 2ab\,(\cos C)$$

$$\cos A = \frac{b^2 + c^2 - a^2}{2bc} \quad \cos B = \frac{a^2 + c^2 - b^2}{2ac} \quad \cos C = \frac{a^2 + b^2 - c^2}{2ab}$$

$$\text{Triangle Area} = 0.5bc\,(\sin A) = 0.5c^2\left(\frac{c^2\,(\sin A)\,(\sin B)}{\sin C}\right) = rs$$

where: $s = 0.5(a + b + c)$

$$r = \sqrt{\frac{(s-a)(s-b)(s-c)}{s}}$$

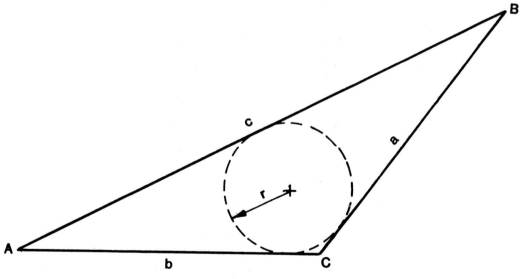

FIG. A-17 Oblique triangle

Arcs. See Figure A-18.

$$\text{Length of Arc } a = \frac{\pi r A^\circ}{180^\circ} = r\rho$$

where: $\rho = \text{rho} = \text{radians}$

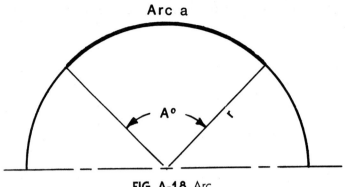

FIG. A-18 Arc

Chords. See Figure A-19.

$$\text{Length of Chord } c = \frac{2r \sin A°}{2}$$

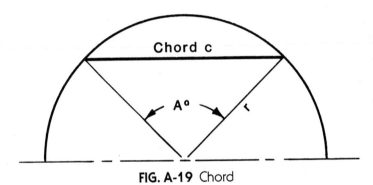

FIG. A-19 Chord

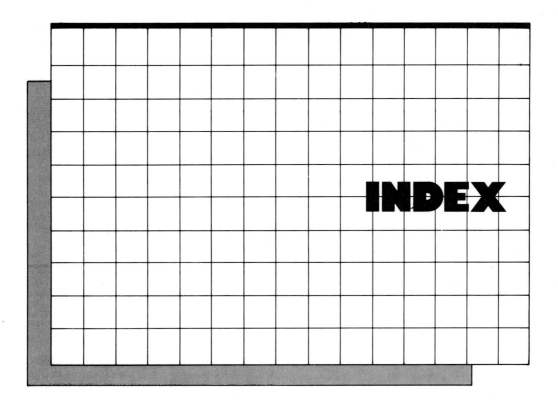

INDEX